PHYSICS AND METALLURGY OF SUPERCONDUCTORS

METALLOVEDENIE, FIZIKO-KHIMIYA I METALLOFIZIKA SVERKHPROVODNIKOV

МЕТАЛЛОВЕДЕНИЕ, ФИЗИКО-ХИМИЯ И МЕТАЛЛОФИЗИКА СВЕРХПРОВОДНИКОВ

PHYSICS AND METALLURGY OF SUPERCONDUCTORS

Proceedings of the Second and Third Conferences on Metallurgy, Physical Chemistry, and Metal Physics of Superconductors held at Moscow in May 1965 and May 1966

Edited by E. M. Savitskii and V. V. Baron

Institute of Metallurgy
Academy of Sciences of the USSR, Moscow

Translated from Russian by G. D. Archard

CONSULTANTS BUREAU · NEW YORK-LONDON · 1970

The original Russian text, first published by Nauka Press in Moscow in
1968, has been corrected by the editors for this edition. The present trans-
lation is published under an agreement with Mezhdunarodnaya Kniga, the
Soviet book export agency.

Е. М. Савицкий, В. В. Барон

Металловедение, физико-химия и металлофизика сверхпроводников

Library of Congress Catalog Card Number 78-107529
ISBN 978-1-4684-8222-5   ISBN 978-1-4684-8220-1 (eBook)
DOI 10.1007/978-1-4684-8220-1

© 1970 Consultants Bureau, New York
Softcover reprint of the hardcover 1st edition 1970

A Division of Plenum Publishing Corporation
227 West 17th Street, New York, N. Y. 10011

United Kingdom edition published by Consultants Bureau, London
A Division of Plenum Publishing Company, Ltd.
Donington House, 30 Norfolk Street, London, W.C. 2, England

PREFACE

The papers collected in this book were presented to the second and third annual conferences on the metallography, physical chemistry, and physics of superconductors which took place in May of 1965 and 1966. These annual conferences, held at the A. A. Baikov Institute of Metallurgy of the Academy of Sciences of the USSR, have quickly become part of the scientific life of the country, and are already a tradition. More than thirty papers were read at each conference, and between 250 and 300 representatives of a large number of organizations were present at each conference.

There was a distinguished array of scientific organizations taking part in the work of these conferences, which discussed current problems in the structure (constitution) and properties of superconductors, and ways of improving their characteristics so as to ensure the successful use of these materials in various new fields of technology.

In the period which has passed since the first conference (May 1964), scientific research into superconducting systems and compounds has undergone substantial further development. A number of diagrams relating the composition to the superconducting properties have been studied; new superconducting alloys have been developed together with methods of processing them and making them into various objects. The phase diagrams of the most promising superconducting systems (Nb—Sn, V—Ga, Nb—Zr, and Nb—Ti) have been investigated more precisely. A start has been made on the study of ternary and more complex alloys. In a number of organizations superconducting magnets and solenoids have been made and work has been carried out on the application of these to various physicotechnical devices and installations.

There has been considerable success in the development of methods of making cable and strip from the superconducting compound Nb_3Sn for the windings of solenoids; magnetic fields exceeding 100 kOe have been achieved in these. In addition to the use of superconducting Nb—Zr alloys in solenoids, Nb—Ti alloys are now being used in windings. The working dimensions of superconducting magnets have increased steadily.

However, the disadvantages retarding the development of superconducting materials mentioned in the first conference still remain. These include the poor stability of the properties of the manufactured materials, their high net cost, the low annual yield, the absence of fully-developed installations for the deposition of shunting and insulating coatings on superconducting wire, and the slow development of methods of constructing superconducting solenoids and making the cryostats and other equipment needed for operations at low temperatures. The commercial output of superconducting wire is insufficiently developed. The training of specialists has lagged.

This cannot fail to affect the tempo characterizing the introduction of superconducting magnets and solenoids into the new technology. Meanwhile the use of such magnetic systems leads in the majority of cases to a sharp reduction in the weight and size of apparatus and devices and to an improvement in their parameters; in some cases systems of an entirely new nature may be made.

To some extent these failings are due to the newness of the problem, which is still only in its initial stage of development.

Our problem is to continue attracting the attention of research workers to the problem of semiconducting materials and to widen and deepen research work in the physical chemistry, metallography, and physics of superconductors. The main problems in this field include: determining the electron structure of superconductors, studying superconducting systems and the structure and properties of compounds and alloys, finding the laws governing changes in the structure and properties of superconductors and discovering new ones, and developing industrial and experimental research into the technology of manufacturing superconducting materials and magnets and using these in apparatus and devices of modern technology.

                                                        E. M. Savitskii and V. V. Baron

CONTENTS

III. THREE-COMPONENT SUPERCONDUCTING ALLOYS

IV. SUPERCONDUCTING PROPERTIES OF COMPOUNDS

AND COATINGS PREPARED FROM THEM

V. PHASE DIAGRAMS OF SUPERCONDUCTING ALLOYS

VI. METHODS OF STUDYING THE PROPERTIES
OF SUPERCONDUCTING ALLOYS

I. PROPERTIES OF NIOBIUM—ZIRCONIUM SUPERCONDUCTING ALLOYS

EFFECT OF IMPURITIES ON THE SUPERCONDUCTING PROPERTIES OF NIOBIUM AND ITS ALLOYS WITH ZIRCONIUM

L. F. Myzenkova, V. V. Baron, and E. M. Savitskii

The effect of interstitial impurities and elements of Groups IVA, VA, and VIA (Zr, Ti, Hf, V, Ta, Cr, W, Mo) on the superconducting properties of niobium is considered. The chief interstitial impurity reducing the value of T_K is oxygen. It is concluded that I_K and H_{c_2} increase with oxygen content. The T_K of niobium changes on alloying with elements of Groups IVA, VA, and VIA. There is a considerable rise in the critical current density of Nb-alloys after introducing oxygen and subjecting to heat treatment; this is attributed to the more effective precipitation of a new phase from the solid solution in the presence of oxygen.

It is well known at the present time that the most favorable combination of superconducting characteristics (critical current density, temperature of the transformation into the superconducting state, critical magnetic field) with technological characteristics is to be found in alloys of niobium and vanadium with Group IVA metals (zirconium, titanium, and hafnium) constituting solid solutions of the substitution type [1].

Niobium—zirconium alloys have received widespread attention; they form a continuous series of solid solutions at high temperatures, and this facilitates the production of superconducting wire, which may be used in widely differing fields of technology.

The aim of this review is to consider the influence of various impurities on the superconducting and technological properties of niobium and its alloys with zirconium.

Niobium

Niobium is a refractory metal of Group VA (melting point 2470°C) with a bcc lattice; it is a superconductor with a transformation temperature of 9.4°K and a critical field between 2.5 and 5 kOe according to various authors.

Table 1 gives the characteristics of niobium and zirconium as regards impurities, the latter depending on the method of production; we see that the amount of gaseous impurities and carbon in niobium varies from hundredths to thousandths of a percent by weight, the main metallic impurity being tantalum. The tantalum content often reaches 0.2 to 0.5 wt.% in electron-beam-melted niobium and in cermet niobium even 2 wt.%; the iron, silicon, and titanium impurities reach tenths of a percent in cermet and hundredths of a percent in electron-beam material, while tungsten and molybdenum impurities reach hundredths to thousandths of a percent. At these concentrations all the metallic impurities dissolve completely in solid niobium. The same table gives

TABLE 1. Typical Impurities in Niobium and Zirconium (wt.%)

| Impurity | Niobium | | Zirconium (iodide type) | Impurity | Niobium | | Zirconium (iodide type) |
	Electron-beam	cermet (metallo-ceramic)			Electron-beam	cermet (metallo-ceramic)	
Ta	0.2—0.5	2	—	Cr	—	—	0.005
Fe	0.01	0.1	0.03	Cd	—	$1 \cdot 10^{-4}$	$3 \cdot 10^{-5}$
Si	0.01	0.09	<0.005	Pb	—	$1 \cdot 10^{-3}$	0.005
Ti	0.01	0.2	0.005	Cu	—	—	<0.005
W	0.02	—	—	Mn	—	—	0.005
Mo	0.02	—	—	O	0.01	0.03	0.05
Hf	—	—	0.01	N	0.01	0.05	0.005
Al	—	—	0.003	H	0.001	0.002	—
Ni	—	—	<0.005	C	0.03		<0.03

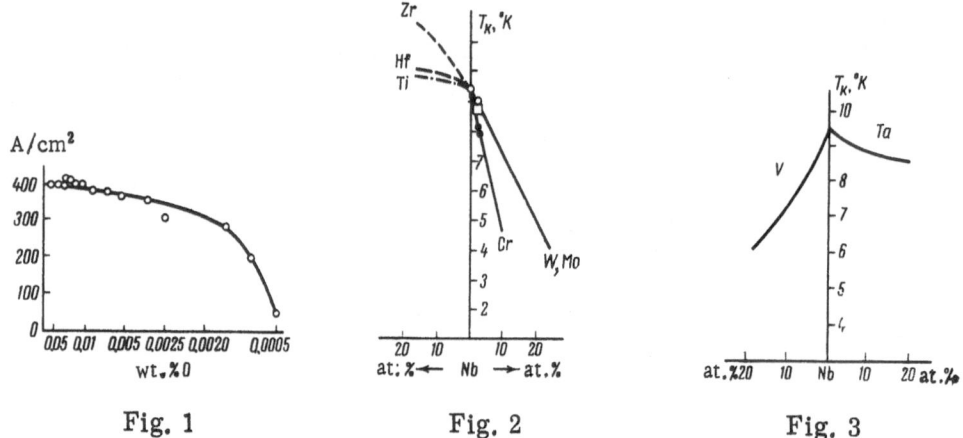

Fig. 1   Fig. 2   Fig. 3

Fig. 1. Critical current density of niobium as a function of oxygen content [10].

Fig. 2. Temperature corresponding to the transformation of niobium into the superconducting state as a function of alloying with Group IVA (Ti, Zr, Hf) and Group VIA (Cr, Mo, W) elements [9].

Fig. 3. Temperature corresponding to the transformation of niobium into the superconducting state as a function of alloying with Group VA (V, Ta) elements [9].

the characteristics of iodide-type zirconium as regards impurities [2]. Zirconium is a metal belonging to Group IVA, with a critical temperature of 0.55°K. We see from Table 1 that the main metallic impurities in zirconium (hafnium and iron) amount to hundredths of a weight percent, while the remaining metallic impurities constitute thousandths or ten-thousandths of a percent; interstitial impurities constitute hundredths of a percent in the case of oxygen and carbon and thousandths in that of nitrogen.

It is well known that interstitial impurities have a strong influence on the ductility of refractory metals. The solubility of interstitial elements in Group VA metals (niobium and tantalum) is 3 to 4 orders higher than in Group VIA metals. In the high state of purity now achieved, niobium constitutes a single-phase solid solution unsaturated with interstitial impurities. As regards their effect on the development of brittleness in VA metals (vanadium, tantalum, and niobium) interstitial impurities may be placed in the following order: hydrogen, nitrogen, oxygen, carbon (the latter have the least effect) [3]. The most harmful impurity is hydrogen; its presence to the extent of 0.002% sharply reduces the ductility of niobium. The

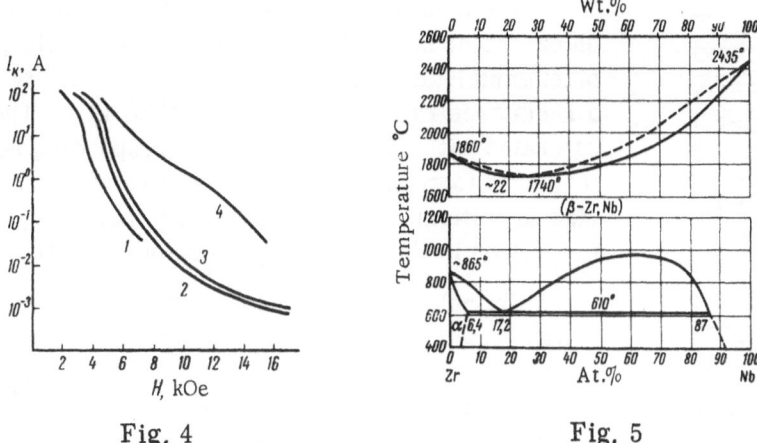

Fig. 4   Fig. 5

Fig. 4. Critical current as a function of the external magnetic
field for the original niobium (1) and the niobium alloyed with 1%
Mo (3), 1% Hf (4) [4].

Fig. 5. Phase diagram of the Nb—Zr system [11].

limits of solubility of the interstitial impurities in niobium at 1200° are: nitrogen 0.014, carbon
0.24, oxygen 0.52%.

The hardness of niobium is a very sensitive indicator of its purity. Thus the hardness of
commercially-pure niobium obtained by vacuum arc melting varies between 100 and 130 kg/mm^2;
electron-beam melting reduces this to 45 kg/mm^2 as a result of purification from interstitial
impurities in the course of melting.

Although niobium has the highest critical temperature and critical field of all the elements, work on the effect of impurities on the superconducting properties started only quite
recently. A number of papers have been devoted to the effect of gaseous impurities on the
superconducting properties of niobium [4-10]. A clear illustration of the effect of a large number of gaseous impurities is given by the critical temperature of the transformation into the
superconducting state: 9.2°K for zone-refined and 5.1°K for powdered niobium [9]. The transformation temperature of niobium obtained by five-fold remelting in an electron beam [6] is
9.46°K. The introduction of oxygen into niobium (within the limits of solubility) reduces the

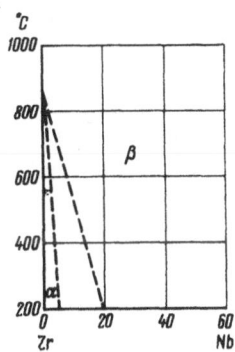

Fig. 6. Phase dia-
gram of the Nb—Zr
system at tempera-
tures below 1000° [14].

transformation temperature to 7.04°K for 0.46 wt.% oxygen (2.6 at.%).
Further increasing the amount of oxygen in the niobium results in
the precipitation of excess oxide phases from the solid solution and
this raises the transformation temperature to 9.02°K for 6.43 at.%
oxygen.

Nitrogen present in niobium to an extent corresponding to the
range of the solid solution reduces the transformation temperature
very slightly, to 9.12°K for 0.33 at.% N; increasing the nitrogen content to concentrations exceeding the solubility limit (1.64 at.%) raises
T_K to 9.24°K.

Hydrogen has a still slighter effect on the transformation temperature of niobium; 3.6 at.% H reduces T_K to 9.22°K.

Thus the main impurity sharply reducing T_K (0.93°K per 1
at.% 0) within the solubility limit is oxygen.

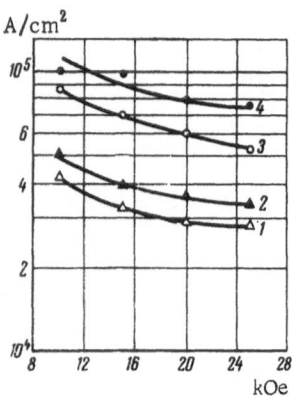

Fig. 7. Critical current density of a Nb–25%Zr alloy as a function of the oxygen and carbon content [21]; 1) without additives; 2) with 0.01% O_2; 3) with 0.02% C; 4) with 0.025% O_2.

The dependence of the critical current density of niobium on oxygen concentration was determined in [10]; the original niobium contained 0.0005 wt.% oxygen (this purity corresponded to a resistance ratio $R_{300°}/R_{4.2°}$, equal to 145). The sample was saturated with oxygen until the corresponding resistance ratio (referred to room and helium temperatures) equalled 8.6. The rise in critical current density with increasing concentration of the oxygen in the niobium was demonstrated convincingly (Fig. 1).

Not only the critical temperature and current, but also the critical magnetic field is extremely sensitive to the amount of oxygen in the niobium [6, 7, 10]. Various structural changes acting on the normal residual electrical resistance or electron specific heat may change the upper critical field, H_{C_2}. It was concluded in [10] that H_{C_2} increasing oxygen content of the niobium, superconductivity being observed in fields exceeding the upper critical field. The critical current density in such fields is also sensitive to oxygen content. The upper critical field of niobium rises to 10 kOe for 1.52 at.% oxygen and 5 kOe for 0.23 at.% nitrogen as compared with 3 kOe for the original niobium [6].

Various binary alloys of niobium with its nearest neighbors in the Mendeleev system (Ti, Zr, Hf, V, Ta, Cr, Mo, and W) were studied in [9]. This work was carried out with zone–refined niobium in which the amount of gaseous impurities (0.01 wt.%) was much lower than that in the niobium powder (0.3 wt.% total) used only a few years before.

Figures 2 and 3 show laws governing the temperature corresponding to the transformation of niobium into the superconducting state on forming binary alloys with Zr, Ti, Hf (Group IVA), V, Ti (Group VA), Cr, W, Mo (Group VIA) [9]. The graphs show a rise in the critical temperature of niobium on introducing IVA elements, a linear fall on introducing VIA elements, and a slower fall on alloying with niobium's neighbors, vanadium and tantalum.

When studying the relation between the transformation temperature of niobium and its content of various other elements in [4], attention was drawn to the dimensional factor, i.e., the difference between the atomic diameter of the solvent and the dissolving elements. It was concluded that T_K rose with increasing concentration of the dissolving elements if their atomic diameter were greater than that of the solvent, but fell if the reverse were the case. The results of [4] and [9] support each other.

Speaking of the relation between the critical temperature of niobium and the alloying additives, it is also interesting to note the effect of these additives on the critical current density. Figure 4 shows the dependence of the current on the external magnetic field both for the original niobium and for the alloyed metal. We see that the addition of 1% hafnium, molybdenum, or tungsten raises the critical current of niobium and increases the critical magnetic field.

Alloys of Niobium with Zirconium

The phase diagram of the Nb–Zr system is shown in Fig. 5. Many authors have studied this phase diagram [11-15]; the majority have drawn the diagram as a continuous series of solid solutions with a minimum on the solidus curve; the decomposition of the bcc cubic solid solution into two bcc solid solutions in combination with the $\alpha \rightarrow \beta$ transformation leads to a monotectoid decomposition. The results of different authors differ slightly in determining the temperature of the monotectoid transformation (615° [11], 800° [13], 560° [12]), the temperature

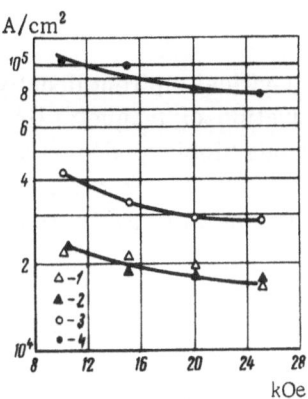

Fig. 8. Critical current density of a Nb–25%Zr alloy as a function of the oxygen content and heat treatment [21]. 1) Alloy cold worked, without oxygen; 2) the same with 0.025% 0; 3) alloy subjected to intermediate annealing at 800°, no oxygen; 4) the same for an alloy with 0.025% 0.

of the onset of decomposition of the β solid solution (1000° [11], 1180° [13]), the solubility of niobium in zirconium, and the monotectoid point. The general form of the diagram is nevertheless confirmed. On working with materials of higher purity a version of the phase diagram of the Nb–Zr system not involving the decomposition $\beta_{s.s.} \rightarrow \beta_{Nb} + \beta_{Zr}$ (Fig. 6) was proposed in [14]. Conversely, on working with metals less free from oxygen, the system contains a three-phase region (typical of three-component diagrams), and the temperature of the monotectoid point moves [15].

The superconductivity of Nb–Zr alloys was discovered in 1952 by Matthias. At the present time alloys of this system have been very widely studied, since they possess high transformation temperatures (maximum T_K 11°K for an Nb–25%Zr alloy), high magnetic fields (about 100 kOe for alloys containing 50 to 60% Zr, and critical currents reaching 10^5 A/cm^2 for an Nb–33%Zr alloy [1, 16, 19].

Work on the effect of impurities on the superconducting and technological properties of niobium–zirconium alloys is very sparse. Evidently the main impurities affecting both kinds of property are gaseous. In view of this a recent paper [20] on the phase state of niobium alloys containing 20 to 25 at.% zirconium and also oxygen is particularly interesting. The limiting solubility of oxygen in these alloys is less than 0.01 wt.% between 1000 and 1200°. For high oxygen contents a second phase separates; the authors identify this as α-zirconium saturated with oxygen. The existence of earlier-proposed zirconium oxides receives no support from x-ray data.

The effect of oxygen and carbon traces on the critical current density of Nb–25%Zr was studied in [21]. Oxygen and carbon were introduced as ZrO$_2$ and ZrC respectively. A wire was prepared in the following way. The original bar was annealed at 1500° for 4 h, it was then cold-forged with 51% reduction, then aged at 800° for 15 min; it was then cold-drawn with a 98.6% reduction. The final diameter of the wire was 0.403 to 0.418 mm. The critical current density was measured in a transverse magnetic field. We see from Fig. 7 that for the alloys containing oxygen and carbon there was a characteristic rise in current density. In order to eliminate the effect of heat treatment, samples containing 0.025 wt.% 0 and other samples entirely free from oxygen were subjected to cold drawing. The field dependence of the critical current density was almost exactly the same for both, as in Fig. 8. A considerable rise in critical current density occurred for the sample containing oxygen after heat treatment. Evidently the rise in critical current density was associated with the decomposition of the solid solution in the presence of oxygen during heat treatment. The authors of the paper in question indicate that, on aging, a metastable phase ZrO (or ZrC in the case of carbon) is formed, its lattice being congruently related to that of the matrix.

Thus by adding reasonable amounts of oxygen, nitrogen, carbon, and hydrogen (keeping within the range of solubility in order to preserve the technological properties of the alloys) the critical current of Nb–Zr alloys may in all probability be increased.

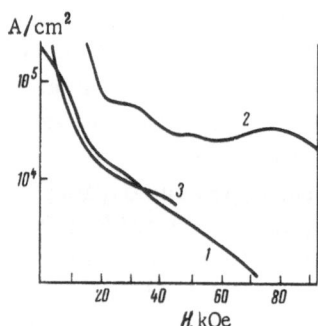

Fig. 9 Critical current density as a function of magnetic field for various niobium alloys [22]: 1) Nb—25%Zr—2%Ta; 2) Nb—25%Zr—5%Ta; 3) Nb—25%Zr—10%Ta.

The main metallic impurity in niobium (and hence Nb—Zr alloys) is tantalum; the proportion of Ta in metalloceramic niobium may reach 2 wt.%. It is well known that tantalum, which forms a continuous series of solid solutions with niobium, does not belong to the set of impurities having an unfavorable effect on the technological properties of Nb—Zr alloys. There is a certain amount of published information [22] regarding the ductility of ternary Nb—Zr—Ta alloys (25% Zr, 2-10% Ta, balance nobium). Alloys with 2 and 10 at.% Ta showed no improvement in superconducting properties although those with 10 at.% Ta were more ductile. The alloy with 5 at.% Ta had a higher current density between fields of 70 and 90 kOe as compared with Nb—Zr alloys (Fig. 9).

The remaining metallic impurities in the original materials (Fe, Ti, Hf) occur as hundredths and thousandths of weight percent and also enter into solid solution; in such quantities these have no serious effect on the superconducting characteristics. It was shown in [23] that iron (not exceeding 0.1 wt.%) in Nb—Zr alloys raised the critical current density slightly, without worsening the machining properties of these alloys. Further raising of the iron content was undesirable, as it led to a reduction in the critical magnetic field (for an Nb—26%Zr alloy) [24].

Conclusions

1. The electron-beam-melted niobium and iodide zirconium now used for producing superconducting alloys satisfy essential requirements, although reducing the interstitial impurities would improve the technological properties of the alloys.

2. Interstitial impurities within the limits at present existing in Nb—Zr alloys raise the critical current density on heat treatment; however, it is undesirable to increase these proportions over a specified limit owing to the sharp worsening of machining properties thereby induced.

3. It is important to obtain niobium and niobium-base alloys of semiconductor purity (millionth parts of impurities) in order to determine the level of their superconducting properties.

LITERATURE CITED

1. E. M. Savitskii and V. V. Baron, Metallurgiya i Gornoe Delo, No. 5, p. 1 (1963).
2. M. A. Filyand and E. I. Semenova, Properties of Rare Elements [in Russian], Izd. "Metallurgiya" (1964), p. 336.
3. E. M. Savitskii, G. S. Burkhanov, and Ch. V. Kopetskii, Izv. Akad. Nauk SSSR, No. 6, p. 12 (1963).
4. W. De Sorbo, Phys. Rev., 130 (6):2177 (1963).
5. W. De Sorbo, Phys. Rev., 132 (1):107 (1963).
6. W. De Sorbo, Phys. Rev., 134 (5A):1190 (1964).
7. W. De Sorbo, Phys. Rev., 135 (5A):1119 (1964).

8. B. T. Matthias, T. H. Geballe, and V. B. Compton, Rev. Mod. Phys., 35 (1):1 (1963).

9. J. K. Hulm and R. D. Blaugher, Phys. Rev., 123 (5):1569 (1963).

10. C. S. Tedmon, R. M. Rose, and J. Wulff, J. Appl. Phys., 36 (1):164 (1965).

11. B. A. Rogers and D. F. Atkins, J. Metals, 7 (9):1034 (1955).

12. Yu. F. Bychkov, A. N. Rozanov, and D. M. Skorov, Atomnaya Énergiya., 2 (2):146 (1957).

13. R. F. Domogala and D. I. McPherson, J. Metals, 2 (5):619 (1956).

14. C. W. Berghout, Phys. Letters, No. 1, p. 292 (1962).

15. H. Richter, P. Wincierz, K. Anderko, and U. Zwiker, J. Less-Com. Metals, 4 (3):252 (1962).

16. T. D. Berlincourt, R. R. Hake, and D. H. Leslie, Phys. Rev. Lett., 6 (12):671 (1961).

17. B. T. Matthias, Phys. Rev., 92 (4):874 (1953).

18. P. R. Aron and H. C. Hitchcock, J. Appl. Phys., 33 (7):2242 (1962).

19. J. Wong, Superconducting Materials [Russian translation], "Mir" (1965), p. 138.

20. V. C. Marcotte, W. L. Larsen, and D. E. Williams, J. Less-Com. Metals, 7 (5):373 (1964).

21. J. O. Betterton, G. D. Kneip, D. S. Easton, and J. O. Scarbrough, Superconducting Materials [Russian translation], "Mir" (1965), p. 102.

22. R. M. Rose and J. Wulff, J. Appl. Phys., 33 (7):2394 (1962).

23. L. F. Myzenkova, V. V. Baron, and E. M. Savitskii, Metallography and Physics of Superconductors [in Russian], Izd. "Nauka" (1965), p. 39.

24. G. B. Kurganov and V. R. Karasik, Metallography and Physics of Superconductors [in Russian], Izd. "Nauka" (1965), p. 118.

EFFECT OF THE ADDITION OF OXYGEN ON
THE STRUCTURE AND SUPERCONDUCTING PROPERTIES
OF ZIRCONIUM—NIOBIUM ALLOYS

Yu. F. Bychkov, I. N. Goncharov, and I. S. Khukhareva

The effect of traces of oxygen on the critical current density of Zr–26-33%Nb alloys is studied. After cold working the critical current of the alloys containing oxygen only slightly exceeds that of oxygen-free alloys, whereas after heat treatment there is a much greater difference. The rise of critical current in alloys containing oxygen is associated with the effect of the precipitation of the O-stabilized α-phase, which differs sharply in superconducting properties from the matrix.

It is well known that the critical current density of hard superconductors is a structure-sensitive characteristic, the value of which is considerably influenced by such metallurgical factors as impurities, the degree and character of deformation, and the presence and distribution of secondary-phase inclusions [1–4]. A study of the influence of oxygen is particularly important because a certain amount of oxygen is inevitably absorbed when preparing and annealing the alloy.

Richter and others [5] first analyzed the problem of the effect of oxygen impurity in the original metals on the form of the Zr–Nb phase diagram. These workers plotted two diagrams, one using spongy zirconium as a base for preparing the alloys and the other with the purer iodide-type zirconium. The spongy zirconium and alloys with 2 and 20% Nb based thereon contained 0.015 wt.% O, while alloys based on iodide zirconium contained about four times less (0.004 wt.%). The diagrams so plotted were considered as sections of the ternary Zr–Nb–O system with a constant oxygen content. A change from iodide to spongy zirconium leads to the broadening of the range of $\beta \rightarrow \alpha + \beta'$ transformations near 18% Nb (monotectoid point) from 100 to 200°. The lower boundary of the range of $\beta \rightarrow \alpha + \beta'$ transformations lies close to 570 or 590°. Richter explains the greater differences in the position of the monotectoid points on the phase diagrams obtained by some authors by differences in the impurity content of the original metals.

An interesting paper by Berghout [6] was also devoted to the effect of oxygen traces on the form of the Zr–Nb diagram. According to this paper the addition of 0.25% O by weight leads to the phase separation of the β solid solutions containing 75% Nb after a 7-h anneal at 800° into two β solid solutions of different concentrations; in alloys without the addition of oxygen no phase separation of the β phase occurs, provided that a layer 0.1 mm thick is removed from the surface before subjecting to x-ray phase analysis. Berghout associates the existence of phase separation on the surface of annealed strip with contamination by oxygen and nitrogen while annealing.

The necessity of allowing for the surface contamination of Zr–Nb alloys with oxygen also follows from work of Slattery [13].

TABLE 1. Presence of the α Phase in Zr–Nb alloys Containing Oxygen
After Annealing at 570°.

Distance from sample surface, mm	26%Nb + wt.% O				33%Nb + wt.% O			
	0	0.18	0.5	1	0.1	0.18	0.25	1
Surface	Much	Much	Much	Much	Much	Much	Much	Much
0.1	None	Very little	Very little	–	Very little	Very little	Very little	–
0.2	–	"	"	–	None	"	"	–
0.3	–	"	"	–	–	"	"	–
0.4	–	"	–	Much	–	–		Much

The effect of oxygen on the properties of Nb–25%Zr was studied in [3]. It was found that the addition of 0.025 wt.% of oxygen (according to analysis the alloy contained 0.05 wt.%) together with intermediate annealing at 800°C led to a considerable rise in the critical current density.

We ourselves also studied the effect of oxygen in Zr–26-33%Nb on the critical current density and the phase separation of the β solid solution. The oxygen was added in quantities up to 0.25 wt.%. The alloy samples were studied in the cold-worked state and after intermediate annealing at 500 to 700°C. It was interesting to study the effect of intermediate annealing on j_K because, on the one hand, the use of intermediate annealing at 400 to 570° enables the value of j_K to be sharply raised in Zr–20-35%Nb not containing oxygen [2], and on the other hand the addition of oxygen considerably affects the decomposition of the β solid solution of Zr–Nb, fundamentally changing the form of the phase diagram [6].

The alloy was prepared in an electric-arc furnace from iodide zirconium and rod niobium. After melting there was only 0.02 to 0.04% of oxygen in the alloys. Bars 55 mm in diameter were forged in air at 800 to 900° into a plate 15 mm thick. After removing the scale the material was cold-rolled to 1 mm. From the resultant strip, plates 1 × 20 × 50 mm in size were cut for saturation with oxygen (of commercial purity); this was done in the apparatus described in [7]. The samples were suspended in a quartz bulb and degassed in vacuum at 800°, weighed, and again placed in the bulb. After introducing oxygen into the bulb in the amount necessary to achieve the specified concentration in the alloy, the sample was heated to between 900 and 930°. The absorption of oxygen was regulated by reference to mercury and oil manometers to an accuracy of 0.5 mm of oil, which constituted less than 1% of the amount of oxygen introduced. In order to secure a uniform distribution of oxygen over the cross section, the strip was vacuum-homogenized at 1300°C in a TVV-4 furnace for 1.5 h. During homogenization additional slight contamination with oxygen took place (Zr–Nb alloys interact mainly with the oxygen rather than the nitrogen in air at 1300°). The additional amount of oxygen absorbed by the samples was determined by weighing on an analytical balance.

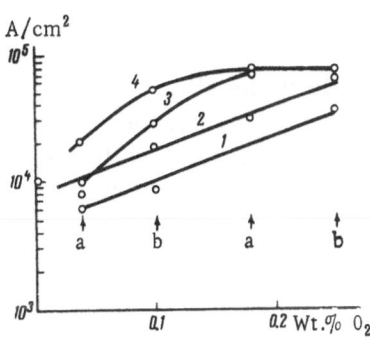

Fig. 1. Critical current density in a field of 30 kOe for zirconium alloys containing 26 (a) and 30 (b) %Nb as a function of the oxygen content. 1) Without intermediate annealing; 2) annealing at 700°; 3) annealing at 570°; 4) annealing at 500°.

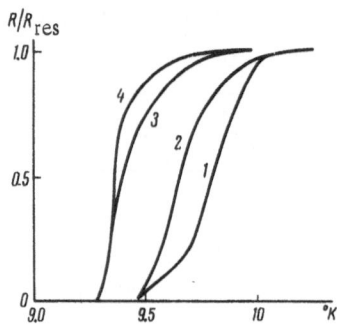

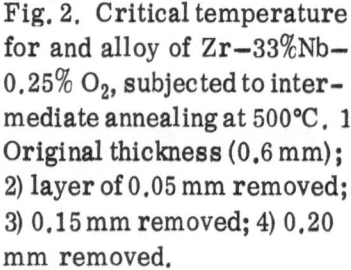

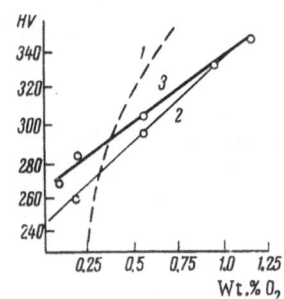

Fig. 2. Critical temperature for and alloy of Zr–33%Nb–0.25% O_2, subjected to intermediate annealing at 500°C. 1) Original thickness (0.6 mm); 2) layer of 0.05 mm removed; 3) 0.15 mm removed; 4) 0.20 mm removed.

Fig. 3. Hardness of the alloys (after homogenization at 1000°) as a function of oxygen concentration. 1) Niobium [8]; 2) Zr–26%Nb; 3) Zr–33%Nb.

The oxygen-saturated, homogenized strips were cold-rolled to 0.5 mm and then subjected to intermediate annealing at 500, 570, and 700° for 1 h, after which they were cold-rolled to 0.05 mm. The critical current density in a magnetic field parallel to the rolling plane was measured in such samples (0.6-mm wide strip). For comparison, strip cold-rolled directly from 1 mm to 0.05 mm without intermediate annealing immediately after homogenization was also tested.

The results of the measurement of j_K for Zr–33%Nb and Zr–26%Nb alloys containing various amounts of oxygen are shown in Fig. 1 for various annealing temperatures. We see that as the oxygen content rises so does j_K. The effect is clearly seen for cold-worked samples and after annealing at 700°C. Annealing at lower temperatures (500 and 570°) leads to the highest absolute values of critical current density, coinciding with those obtained for an alloy free from oxygen [12]; however, in the case of the alloys containing oxygen the relative rise in j_K as a result of heat treatment was considerably less.

One notices the fact that after intermediate annealing at 700°C the value of j_K for the alloy without oxygen falls almost to the value of the cold-worked alloy, while in the alloy with a high oxygen content (0.25 wt.%) j_K remains quite high. We note that the degree of deformation for these samples is different; thus, for samples without oxygen the reduction is 200:1 while for the alloy containing oxygen (as a result of the homogenization of sheets 1 mm thick) it is reduced to 17:1.

We see from Fig. 1 that the relation between j_K and oxygen concentration changes in character with the temperature of intermediate annealing. It is clear that the reason for the changes in the j_K of alloys containing oxygen should be sought in the structural characteristics of these alloys.

Metallographic study of a Zr–33%Nb–0.18%O alloy after homogenization at 1300° showed that it constituted a solid solution with a small quantity of second-phase inclusions $3 \cdot 10^{-4}$ cm in size. After annealing at 500 to 700° all the samples with oxygen content were two-phased. The effect of oxygen on the decomposition of the β solid solution in the interior of the samples was determined by an x-ray method after annealing at 570°, i.e., below the monotectoid temperature. The decomposition of the β solid solution was detected by reference to the presence of the α phase (see Table 1).

In the original alloy (unsaturated by oxygen) all three phases occur on the surface: the α phase, the original β phase, and the niobium-rich β' phase. At a depth of 0.1 mm the lines of the α phase vanish. In alloys with 0.18 and 0.5% oxygen the α phase appears even at a depth of 0.4 mm, although only in small quantities. In alloys containing 1% oxygen the amount of α phase remains constant in depth and is quite high.

The alpha phase in the oxygen-saturated samples is formed as a result of the stabilization of the zirconium [11], i.e., the oxygen mainly combines with the zirconium, forming a Zr–O solid solution. Since this phase is also present in samples not subjected to intermediate annealing, it may clearly be associated with the rise in the j_K of the oxygen-saturated alloys. According to Anderson [9] this kind of heterophase structure (the presence of α-Zr inclusions, which are not superconducting at 4.2°K, in a superconducting matrix) should favor the retention of a high critical current density in strong magnetic fields.

The measurement of T_K for a Zr–33%Nb–0.25%O$_2$ alloy subjected to various kinds of annealing showed that the amount of niobium-enriched β' phase (the phase with the high value of T_K) responsible for the rise in T_K on annealing gradually diminished on passing deeper into the sample. We see from Fig. 2 that the vanishing of the niobium-rich β' phase leads to a fall in T_K after removing the surface layers.

No changes in the critical current density took place after removing the surface layers subject to decomposition on heat treatment (0.1 to 0.2 mm thick); i.e., the change in j_K was not associated with appearance of a phase having a higher T_K [2].

It follows from this that the presence of a phase with a higher T_K does not lead to a rise in j_K at 30 kOe; on the other hand the presence of α-Zr does cause a rise in j_K. An increase in the oxygen content of the alloys raises their hardness (Fig. 3) after homogenization at 1300°C. However, the rise in hardness is smaller than in the case of pure niobium and niobium-base alloys [8].

After annealing at 700°C the zirconium alloy containing 33% Nb and 0.25% O softened very little, while the hardness of the alloy containing 26% Nb and 0.04% O fell by 20 units.

Conclusions

1. We may consider that the rise in the j_K of alloys containing oxygen takes place as a result of the presence of fine precipitated particles of α-Zr in the superconducting matrix; these have poor superconducting parameters and exert a stabilizing effect on the magnetic flux lines [9, 10].

2. The use of oxygen-saturated original materials leads to a rise in the critical current density but also to a fall in ductility.

3. Zirconium-base alloys allow the introduction of larger amounts of oxygen than pure niobium and niobium-base alloys, owing to the smaller influence of oxygen on the ductility of these alloys.

Large amounts of oxygen favor the decomposition of the β solid solution with the precipitation of the oxygen-stabilized α phase.

LITERATURE CITED

1.   V. D. Borodich, A. P. Golub', et al., Zh. Éksp. Teor. Fiz., 44:110 (1963).
2.   Yu. F. Bychkov, I. N. Goncharov, V. I. Kuz'min, and I. S. Khukhareva, Pribory i Tekh. Éksperim., No. 3, p. 170 (1964).
3.   I. O. Betterton, G. D. Kneip, D. S. Easton, and I. O. Scarbrough, Superconductors, New York, Interscience (1962), p. 61.

4. R. G. Trenting et al., High Magnetic Fields, New York (1962).
5. A. C. Richter, J. Less-Com. Metals, 4 (3):252 (1962).
6. C. W. Berghout, Phys. Letters, July, No. 1, p. 292 (1962).
7. N. V. Borkov, in: Metallurgy and Metallography of Pure Metals, No. II [in Russian], Atomizdat (1960), p. 148.
8. D. A. Prokoshkin and E. V. Vasil'eva, Alloys of Niobium [in Russian], Izd. "Nauka" (1964).
9. P. E. Anderson, Phys. Rev. Letters, No. 9, p. 309 (1962).
10. P. W. Anderson and Y. B. Kim, Rev. Mod. Physics, 36:39 (1964).
11. V. C. Marcotte, W. L. Larsen, and D. E. Williams, J. Less-Com. Metals, 7 (5):373 (1964).
12. Yu. F. Bychkov, I. N. Goncharov, and I. S. Khukhareva, Zh. Éksp. Teor. Fiz., 48:818 (1965).
13. G. F. Slattery, J. Less-Com. Metals, 8 (3):195 (1965).

X-RAY DIFFRACTION AND METALLOGRAPHIC STUDY
OF PHASE TRANSFORMATIONS IN SUPERCONDUCTING
NIOBIUM—ZIRCONIUM ALLOYS

N. F. Pravdyuk, G. P. Saenko, and L. A. Elesin

The phase composition and structure of cold-worked and annealed (at 500, 700, and 1000°) Nb-30-50% Zr alloys are studied by x-ray methods and in the electron microscope. As a result of a decomposition at 500° an Nb-base bcc solid solution and a solid solution of Nb in α-Zr with a hexagonal lattice are formed. Annealing at 700° causes decomposition into two bcc solid solutions. Annealing at 1000° causes recrystallization of the alloys, taking place most completely in that containing 50% Zr.

The purpose of this investigation was to determine the nature of the phases formed in superconducting Nb—Zr alloys at various annealing temperatures, their size, shape, and distribution. The literature [1, 2] tends to contain information simply relating to the final results of the decomposition of solid solutions of the Nb—Zr system, i.e., the situation after the alloys have passed into an equilibrium state. There are no data regarding the kinetics of the transformations and the formation of metastable phases (size and distribution). However, in the light of modern theoretical views regarding superconductors of the second kind [3], it is precisely the microscopic inhomogeneities and structural imperfections (resulting from alloying and heat treatment) which determine high superconducting current density and high critical fields. An all-round study of the structure of the alloy and structural changes arising from various causes (in combination with the measurement of superconducting properties) is important in order to establish the optimum treatment of the alloy for obtaining maximum critical properties and also for understanding the mechanism of superconductivity itself.

The original material for the investigation was cold-rolled strip obtained from NTs30 and NTs50 alloys (Nb—30%Zr and Nb—50%Zr), subjected to double electron-beam remelting. After melting, the bars were subjected to a homogenizing anneal at 1600°C. The degree of reduction of the strip rolled from the bars was 96%. The samples cut from the strip were annealed at 500, 700, and 1000°C for various periods (15 min to 5 h). The annealing took place in a quartz ampoule at a residual-gas pressure of $2 \cdot 10^{-5}$ and $2 \cdot 10^{-6}$ mm Hg. The accuracy of temperature control was ±5°.

The x-ray diffraction pictures were taken in standard Debye cameras of the RKU-86 type in Co K_α-radiation, using a 20-μ iron foil as filter. The sample (20 × 0.2 × 0.1 mm in size) was rotated while the photograph was taken. Samples of two sorts were studied; these were respectively cut along and across the strip-rolling direction. For phase analysis of the alloys in the original state and also after annealing at 500 and 700°C x-ray diffraction pictures from transverse samples were used (these had a large number of lines and more even photometric density). For alloys annealed at 1000°C the large number of lines occurred in the case of the longitudinal samples.

TABLE 1. Annealing of Cold-Worked NTs50 at 700°C

Annealing period, h	Pressure, mm Hg	Observed phases (crystal structure and lattice constant in Å)
0.25	$2 \cdot 10^{-6}$	1. bcc $(a = 3.500 \pm 0.003)$ 2. bcc $(a = 3.449 \pm 0.006)$ 3. bcc $(a = 3.334 \pm 0.003)$
0.5	$2 \cdot 10^{-5}$	1. bcc $(a = 3.499 \pm 0.003)$ 2. bcc $(a = 3.447 \pm 0.005)$ 3. bcc $(a = 3.333 \pm 0.005)$ 4. hex $(a = 3.24; c = 5.17)$
1	$2 \cdot 10^{-6}$	1. bcc $(a = 3.502 \pm 0.003)$ 2. bcc $(a = 3.450 \pm 0.005)$ 3. bcc $(a = 3.325 \pm 0.005)$
3	$2 \cdot 10^{-5}$	1. bcc $(a = 3.500 \pm 0.003)$ 2. bcc $(a = 3.332 \pm 0.003)$ 3. hex $(a = 3.24; c = 5.17)$
5	$2 \cdot 10^{-6}$	1. bcc $(a = 3.502 \pm 0.001)$ 2. bcc $(a = 3.325 \pm 0.001)$

Metallographic examination was carried out by the replica method in an electron microscope. The microstructure of the alloys was studied in the rolling plane and also in perpendicular planes along and across the rolling direction. After electropolishing, the samples were subjected to cathodic vacuum etching in a UVR-2 apparatus in an atmosphere of spectrally-pure neon with a cathode voltage of 5 kV and a current of 5 μA. The etching time was about 30 min. After completing the etching, carbon replicas were obtained from the microsections (shadowed with uranium); these were inspected in the electron microscope at a voltage of 80 kV.

Phase Composition of Cold-Worked Alloys After

Annealing at 500, 700, and 1000°C

The x-ray diffraction pictures of alloys subjected to various forms of heat treatment were analyzed in the usual way. At the first stage the interplane distances were determined for all the visible lines on the diffraction picture. Then series of lines relating to different phases were separated out from the whole set of d_{HKL} values. The lines were indexed analytically and by comparing with known substances.

Alloy NTs50. The x-ray diffraction pictures of the longitudinal and transverse samples of the cold-worked alloy showed seven diffuse lines relating to a phase with a bcc lattice and parameter 3.433 ± 0.003 Å. According to the curve relating the lattice parameter of Nb–Zr alloys to niobium concentration [1], the phase in question was a solid solution of niobium and zirconium of equiatomic composition.

A three-hour anneal of the cold-worked samples at 500°C and a pressure of $2 \cdot 10^{-5}$ mm Hg led to the appearance of additional lines on the x-ray pictures. On analysing these pictures, all the lines were separated into three series. The first series (seven lines) was analogous to that observed in the x-ray diffraction pictures of the cold-worked material, i.e., it corresponded to the original solid solution. The intensity of the four lines of the second series was much weaker; these lines related to another phase with a bcc lattice but a parameter 3.340 ± 0.005 Å. The remaining five lines belonged to the third series; as regards their positions, these were similar to certain lines observed in the x-ray picture of pure zirconium taken under analogous conditions. The phase represented by the third series of lines evidently had a hexagonal crystal lattice with parameters a = 3.24 and c = 5.15 Å.

TABLE 2. Phase Composition of Cold-Worked NTs50 After a Three-Hour Anneal
at Various Temperatures in a Vacuum of $2 \cdot 10^{-5}$ mm Hg

Annealing temp., °C	Structure and nature of phases observed	Source of phase formation
Unannealed alloy	bcc, a = 3.433 Å, solid solution Nb–50%Zr	Original solid solution
500	1. bcc, a = 3.435 Å, solid solution Nb–50%Zr 2. bcc, a = 3.340 Å, solid solution Nb–18%Zr 3. hex, a = 3.24 Å, c = 5.15 Å, solid solution of Nb in α-Zr	Nondecomposed original solid solution Decomposition products of original solid solution
700	1. bcc, a = 3.500 Å, solid solution Nb–50%Zr 2. bcc, a = 3.332 Å, solid solution Nb–12%Zr 3. bcc, a = 3.45 Å, solid solution Nb–55%Zr 4. hex, a = 3.24 Å, c = 5.17 Å, solid solution of gases in α–Zr*	Decomposition products of original solid solution Original solid solution Product of interaction with gases
1000	1. bcc, a = 3.433 Å, solid solution Nb–50%Zr 2. fcc, a = 4.634 Å, mixed carbide (*)	Recrystallized original solid solution Product of interaction with gases

*Proposed phase

The x-ray diffraction photographs of annealed samples subjected to electropolishing in order to remove a surface layer 20 μ thick showed that the new cubic and hexagonal phases were not surface effects. The appearance of these phases at 500°C agrees qualitatively with the known phase diagram of the Nb–Zr system [1, 2] according to which the solid solution of niobium with zirconium should decompose at 500°C with the formation of two solid solutions based on niobium (bcc lattice) and α-zirconium (hexagonal). After three hours annealing at 500°C only a small proportion of decomposition products is formed, the greater part of the solid solution remaining undecomposed. By analyzing the line intensities one finds that the volume of phases precipitating is no greater than 5% of the volume of the whole material. In addition to this, in contrast to the composition in a state of equilibrium, the Nb-base solid solution forming contains not 12 but 18 at.% Zr. The crystals of the decomposition products and the original phase show preferred orientation.

As a result of a three-hour anneal at 700°C in a vacuum of $2 \cdot 10^{-5}$ mm Hg the corresponding x-ray pictures of NTs50 showed a considerable number of interference maxima. Two series of lines characterized phases with bcc lattices: The first of these (seven diffuse lines of medium intensity) related to a phase with a lattice parameter of 3.500 ± 0.003 Å, the second series (six diffuse lines of high intensity) defined a phase with a parameter of 3.332 ± 0.003 Å. The third series of lines (15 medium-intensity maxima) agreed closely with those of zirconium annealed under similar conditions; the lattice parameters of this phase were a = 3.24 Å, c = 5.17 Å; the hexagonal phase was less distorted than the first two (sharper interference maxima, resolved K_{α_1}-K_{α_2} doublet). All three phases possess preferred orientation.

Electropolishing of the samples showed that the hexagonal phase was a surface phase and extended only about 30 μ down (the total thickness of the sample was some 100 μ). In the deeper layers only the two cubic phases remained, these constituting the product of the decomposition of the original solid solution. One phase (a = 3.500 Å) was a solid solution of niobium (about 30 at.%) in β-zirconium and the other (a = 3.332 Å) was a solid solution of zirconium

Fig. 1. Microstructure of cold-worked NTs50 (× 18,000, reduced 50% for reproduction.

(about 12 at.%) in niobium. The intensity of the corresponding lines of the cubic phases meanwhile became identical. This indicated that the hexagonal phase in the surface layer grew at the expense of a reduction in the amount of the Zr-rich cubic phase. The x-ray pictures of the polished samples showed two or three weak lines of the original solid solution; however, the lattice constant of this (3.450 Å) was somewhat larger than in the unannealed state (3.433 Å).

Table 1 shows results obtained for samples of NTs50 annealed at 700°C for various periods and residual-gas pressures. We see that the hexagonal phase is present when the gas pressure in the system equals $2 \cdot 10^{-5}$ mm Hg; on reducing the pressure by one order the lines of this phase vanish almost entirely, or at any rate only traces of the strongest lines of the series appear. It would seem that the hexagonal phase appears as a result of the interaction of residual gases in the vacuum system with the surface of the heated metal and that it constitutes a solid solution of the gaseous elements in α-zirconim.

The dimensions of the particles of the precipitating phases increase gradually with annealing period (so does the number of particles), so that after a three-hour anneal only a small amount of the original solid solution remains. A five-hour anneal leads to complete decomposition; the increased sharpness of the lines and the incipient separation of the doublet indicate that the size of the particles in the phases being formed has become greater than 10^{-5} cm. It is interesting to note that, in the course of the transformation, the solid solution still in the original phase is nevertheless not entirely unaltered. This is indicated by the increased lattice constant (3.43 to 3.45 Å), evidently attributable to the loss of some of the niobium atoms. T .e type of decomposition products corresponds to the phase diagram of the Nb–Zr system, although the composition of the new phases fails to agree, evidently because of the incompletion of the transformation process. As in the case of annealing at 500°C, the precipitating crystals are preferentially oriented in the sample.

According to the x-ray data, annealing the alloy NTs50 at 1000°C leads to the removal of stresses and to the recrystallization of the original alloy without decomposition. On the surface of the sample, interaction with gases from the vacuum system produces a phase with an fcc lattice. Analysis of the substances formed by the annealing of pure zirconium shows that the phase with the fcc lattice constitutes mixed niobium and zirconium carbides.

The results of the phase analysis of NTs50 are given in Table 2.

Alloy NTs30. The x-ray-diffraction pictures of cold-worked samples show seven diffuse lines relating to a phase with a bcc lattice and a parameter of 3.375 Å. Such a lattice corresponds to a solid solution of about 30 at.% Zr in niobium.

Annealing cold-worked samples at 500°C for 3 h produced no visible changes in the x-ray diffraction pictures as compared with the original state.

However, after heating at 700°C the diffraction picture changes. The x-ray diffraction pictures obtained indicate that the annealing of NTs30 at this temperature leads to the decomposition of the worked solid solution, with the formation of two new cubic phases: a solid solution

TABLE 3. Phase Composition of Cold-Worked NTs30 After Three-Hour Annealing at Various Temperatures in a Vacuum of $2 \cdot 10^{-5}$ mm Hg

Annealing temp °C	Structure and nature of observed phases	Source of phase formation
Unannealed alloy	bcc, a = 3.375 Å, solid solution of Nb–30%Zr	Original solid solution
500	bcc, a = 3.375 Å, solid solution of Nb–30%Zr	Nondecomposed original solid solution *
700	1. bcc, a = 3.499 Å, solid solution of Nb–70%Zr 2. bcc, a = 3.323 Å, solid solution of Nb–13%Zr 3. bcc, a = 3.350 Å, solid solution of Nb–22%Zr 4. hex, a = 3.24 Å, c = 317 Å, solid solution of gaseous elements in α-Zr †	Decomposition products of original solid solution Original solid solution Products of interaction with gases
1000	1. bcc, a = 3.367 Å, solid solution of Nb–30%Zr 2. fcc, a = 4.634 Å, mixed carbide of Nb and Zr †	Recrystallized original solid solution Products of interaction with gases

*Analysis of the microstructure in the electron microscope showed that decomposition in fact occurred; however, the amount of decomposition products fell below the sensitivity of the x-ray method.
†Proposed phase

of niobium (about 30 at.%) in β-zirconium (a = 3.499 Å), and a solid solution of zirconium (about 13 at.%) in niobium (a = 3.323 Å). After a three-hour anneal a considerable amount (up to 50 vol.%) of undecomposed original solid solution still remains.

However, the crystal lattice parameter of the latter falls slightly (from 3.375 to 3.350 Å) as a result of the loss of some of the zirconium atoms (up to 22 at.%). On the surface of the samples a hexagonal phase is formed (as in the case of NTs50), owing its generation to the interaction of the heated metal with gases.

The lines of the x-ray pictures of the alloy annealed at 1000°C (3 h, $2 \cdot 10^{-5}$ mm Hg) show that the alloy contains phases with a bcc lattice (3.367 ± 0.003 Å) and with an fcc lattice (4.634 ± 0.005 Å). The lines of the first phase are slightly diffuse. The doublets are not resolved. Hence we may conclude that after a three-hour anneal at 1000°C relaxation and partial recrystallization processes occur in the original solid solution.

On the surface of the sample a product formed by the interaction of the alloy with residual gases of the vacuum system appears, evidently constituting a mixed carbide of zirconium and niobium. The results of the phase analysis of the alloy NTs30 are given in Table 3.

Study of the Microstructure of the Alloys

The structure of the single-phase 96% cold-worked NTs30 and NTs50 alloys is identical. By way of example Fig. 1 shows a photograph of the microstructure of worked NTs50 (in the rolling plane). We notice parts of the grains drawn out along the rolling direction, broken into fragments.

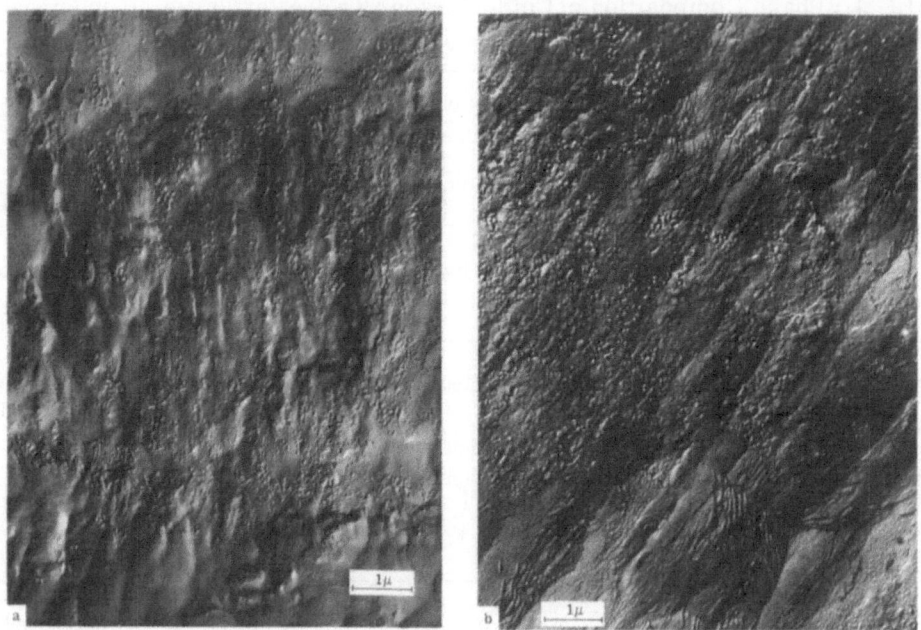

Fig. 2. Microstructure of cold-worked alloys NTs50 (a) and NTs30 (b) after annealing at 500° for 3 h (× 18,000, reduced 50% for reproduction).

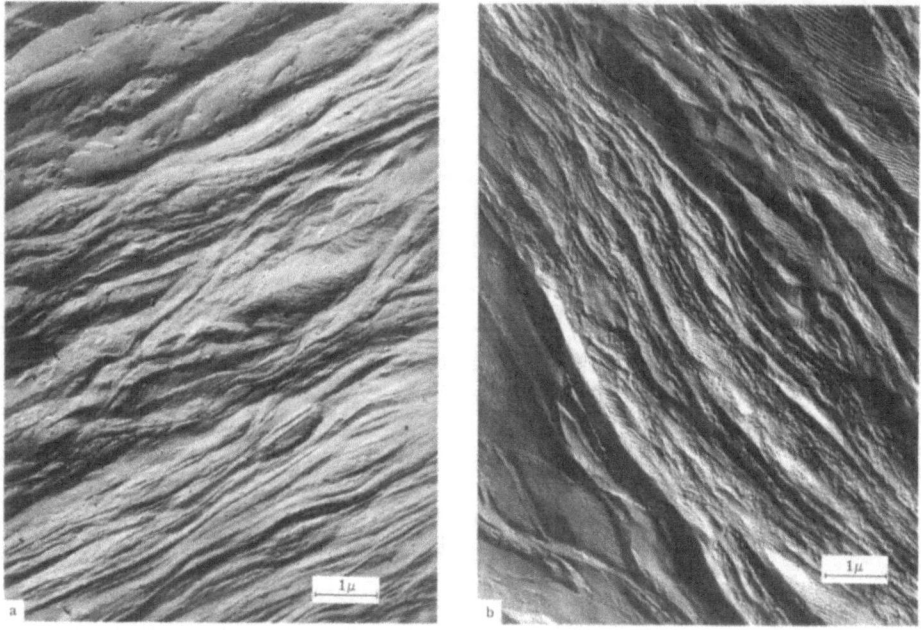

Fig. 3. Microstructure of cold-worked NTs50, perpendicular to the rolling direction (× 18,000, reduced 50% for reproduction). a) Before annealing; b) after annealing.

During the 500° anneal both alloys undergo decomposition. The x-ray pictures only show traces of the transformation in the case of NTs50, while in NTs30 (also annealed for 3 h at 500°) no additional lines appear. However, the crystallites of the precipitating phases are easily visible in both cases on analyzing the structure in the electron microscope (Fig. 2).

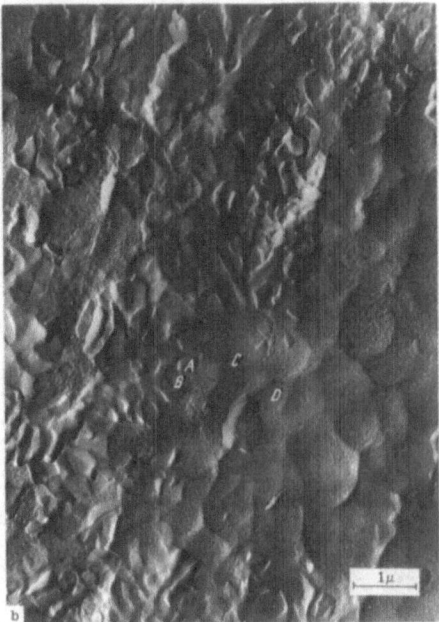

Fig. 4. Microstructure of cold-worked alloy NTs50 after annealing at
700° for 3 h (× 18,000, reduced 50% for reproduction). a) Part showing complete decomposition; b) part showing partial decomposition.

The precipitating phases constitute particles of rounded or elongated form, distributed
over the matrix of the original solid solution. The particle size varies from 100 to 1000 Å. The
particles are distributed in a nonuniform manner, forming aggregates of varying shapes and
sizes.

The size and number of the precipitating crystallites indicate that the transformation only
takes place partially in the alloys, the greater part of the volume of the solid solution not de-
composing. In addition to this, the parts of structure shown in Fig. 2. are not characteristic
of the whole sample; on inspection in the electron microscope, regions with no trace of decomposition may also be seen. Although the size of the particles of precipitating phases is almost
the same in both alloys, the total amount is much smaller in NTs30 than in NTs50. As a result
of this the x-ray pictures of the 500°-annealed NTs30 show no lines attributable to decomposition products.

Figure 3 shows the distribution of the precipitating phases with respect ot the thickness
of a strip of NTs50. The particles here form aggregates in the shape of sinuous bands, alternating with nondecomposed parts of the solid solution. In the main volume of the alloy the precipit-
ate zones evidently have the form of sinuous layers up to 0.3 μ thick and less than or equal to
the width of the distorted solid-solution grains in width.

A three-hour anneal of the worked alloys at 700°C caused decomposition over a much greater volume of material than at 500°C. However, even in this case the relative amount of decomposition products in NTs30 was less than in NTs50. The decomposition processes occurred
nonuniformly at 700° also.

Figure 4 shows the microstructure of two parts of a sample of NTs50 along the rolling
plane. In that part of the material in which decomposition has occurred practically completely
(Fig. 4a) the average size of the approximately equiaxial decomposition products is about 0.25
μ (fluctuating from 0.1 to 0.5 μ). In those parts in which only part of the material has been em-
braced by the transformation (Fig. 4b) the average grain size is rather larger. The same figure

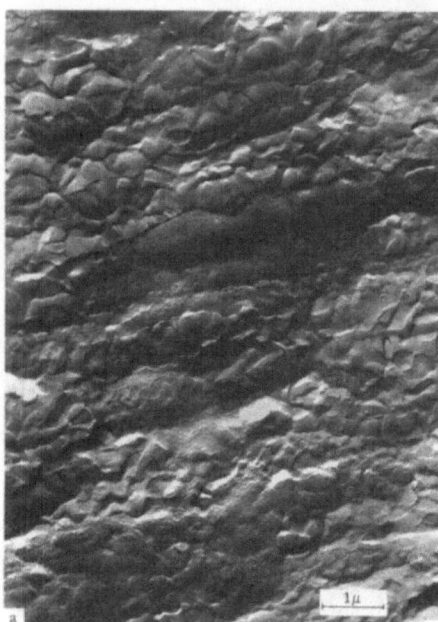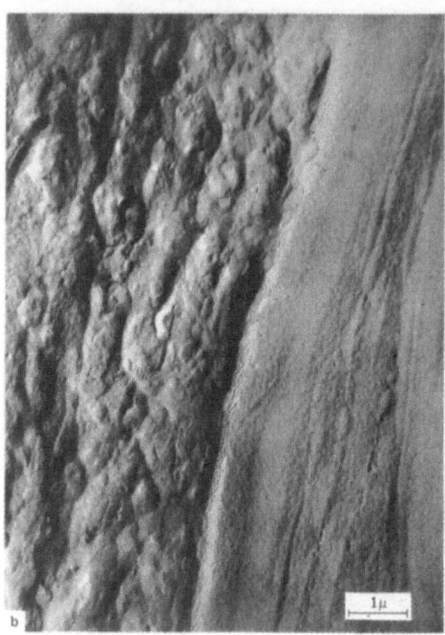

Fig. 5. Microstructures of cold-worked alloys NTs30 and NTs50 after annealing at 700° for 3 h (× 18,000, reduced 50% for reproduction). a) Alloy NTs30, in rolling direction; b) alloy NTs50, perpendicular to rolling direction.

clearly shows the two-phase character of the decomposition of the original solid solution. For example, the elements A and B of the structure are crystals of the two precipitating cubic phases. We also clearly see parts of the solid solution in which crystals of the new phases have been generated (C) and other regions outwardly remaining unaltered (D). It was found by x-ray diffraction that in reality the regions of undecomposed solid solution slightly changed their composition (see Tables 2 and 3). Figure 5a shows an electron micrograph of the structure of NTs30 in the rolling direction; we see that the outward form of the precipitating particles and their average diameter are roughly the same as in NTs50. As already noted, the volume of the nondecomposed solid solution in NTs30 is larger than in NTs50. In order to calculate the volume of the precipitating particles of the new phases, a series of micrographs was obtained from NTs30 and NTs50 samples with a minimum magnification (3500×), the precipitates being quite clearly seen in this way. In the parts of the samples studied this volume equals up to 80% (NTs50) and 50% (NTs30) of the total amount of material.

The distribution of the decomposition products over the thickness of the sample may be seen from Fig. 5b, in which the microstructure of NTs50 is shown perpendicular to the rolling direction. The figure shows contiguous parts of two grains from a region of the original solid solution which has undergone transformation. The distribution is in this case more uniform than after annealing at 500°C.

In order to study the structure of superconducting alloys, a method of studying thin films by transillumination has been developed; this enables various crystallographical calculations to be carried out and changes in both the phase composition and dislocation structure of the material to be followed. Some experiments in this direction with NTs30 and NTs50 (which are difficult to thin) have already been carried out. By way of example Fig. 6 shows the structure of NTs50 after 96% cold working and annealing at 700° for 3 h. We notice the grains of the precipitating bcc phases; owing to the different orientation the contrast of the grains varies over wide limits. The mean grain size is $0.2\,\mu$. In a number of grains (A) dislocations may be seen. The micrograph also shows parts in which the decomposition of the solid solution is only beginning (region B).

Fig. 6. Microstructure of cold-worked alloy NTs50 in the rolling plane after annealing at 700°. Thinning method. (× 75,000, reduced 50% for reproduction.)

The three-hour annealing of cold-worked NTs50 and NTs30 leads to recrystallization, the composition of the original solid solution being preserved. The crystallization process takes place more completely in NTs50 (Fig. 7a); the size of the polyhedral grains formed equals 20 μ. In NTs30 (Fig. 7b) the size of the largest grains is only half this.

Conclusions

The annealing of cold-worked superconducting alloys NTs30 and NTs50 at 500 and 700°C leads to the decomposition of the original solid solution.

As a result of the decomposition at 500°C two new phases are formed in the alloys: a solid solution of zirconium (18 at.%) in niobium with a bcc lattice and a solid solution of niobium in α-zirconium with a hexagonal lattice. After a three-hour heating the amount of decomposed original solid solution in NTs50 is roughly 5% of the original volume, and still less in NTs30. The particle size of the precipitating phases lies between 100 and 1000 Å. Within the alloy these particles form aggregates of various shapes, alternating with parts of the alloy in which no decomposition has taken place.

On heating to 700° two phases with bcc lattices separate out from the solid solution of both alloys: a solid solution of niobium (about 30 at.%) in β-zirconium and a solid solution of zirconium (about 12 at.%) in niobium. After holding for three hours at 700°, up to 80% of the original solid solution decomposes in NTs50 and up to 50% in NTs30. The mean grain size of the precipitating phases equals 0.25 μ (from 0.1 to 0.5 μ). The decomposition is characterized by a considerable nonuniformity in this case also; however, the decomposition products are spread more uniformly through the thickness of the strip than after annealing at 500°. The solid solution not undergoing decomposition at 700° changes its composition; in NTs50 about 5% Nb is lost and in NTs30 about 8% Zr. A study of the kinetics of the decomposition of NTs50 showed that the change in the composition of the original solid solution was completed after a 15-min holding period, and subsequently there was only a reduction in the amount of this phase present as the decomposition process continued. The decomposition into two phases in NTs50 was completed in 5 h at 700°.

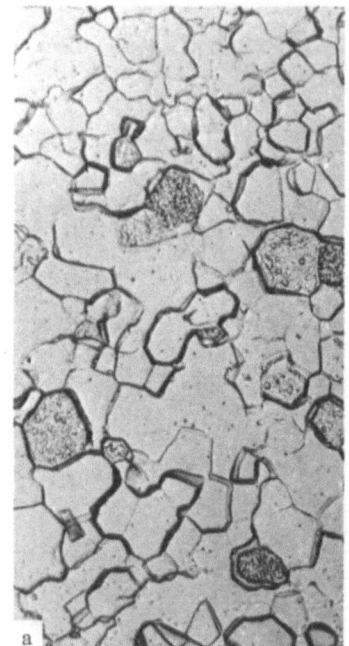 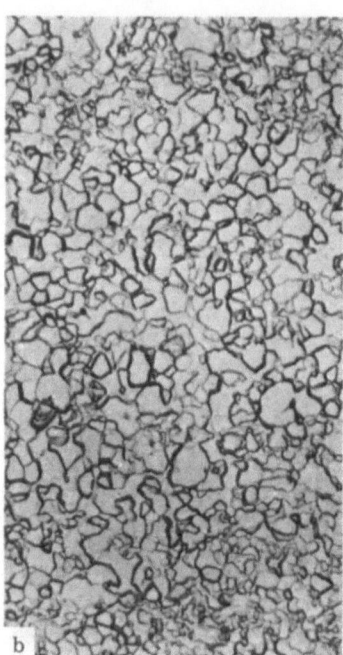

Fig. 7. Microstructure of NTs50 (a) and NTs30 (b) alloys after annealing at 1000° for 3 h (× 300).

Annealing NTs30 and NTs50 at 1000° causes no decomposition of the original solid solution. Recrystallization takes place instead. The recrystallization process takes place most
completely in NTs50, the grain growing to 20 μ. In NTs30 the maximum size of the growing
grain is 10 μ.

On heating the alloys in vacuum (10^{-5} mm Hg) to 700° a phase with a hexagonal lattice is
formed in the surface layer to a depth of 30 μ. This phase appears as a result of the interaction of the alloys with residual gases in the vacuum system and increases at the expense of a
reduction in the amount of one of the cubic phases (decomposition products) rich in zirconium.
It may be considered that the surface phase is a solid solution of gaseous elements in α-zir-
conium. Heating to 1000° in the same vacuum causes the formation of a phase with an fcc lattice
in the surface layer. Analytical results suggest that this phase is a mixed carbide of niobium
and zirconium.

On the basis of the experimental results obtained it may be considered that the precipitation of very fine particles (uniformly distributed with respect to volume) from the solid solutions of NTs30 and NTs50 will facilitate the creation of higher, linearly-uniform critical properties in superconducting wires made from these alloys.

LITERATURE CITED

1.   R. Rogers and D. Atkins, J. Metals, 7 (9):1034 (1955).
2.   Yu. F. Bychkov, A. N. Rozanov, and D. M. Skorov, Atomnaya Énergiya, 2 (2):146 (1957).
3.   C. B. Beam and R. W. Schmitt, Science, 140 (3562):26 (1963).

PROPERTIES OF AN ALLOY WITH
A DISPERSED SUPERCONDUCTING PHASE

E. P. Romanov, V. D. Sadovskii,
N. V. Volkenshtein, and L. V. Smirnov

The effect of the conditions of the decomposition of the supersaturated solid solution in a Zr—4%Nb alloy on the value of T_K is considered. The effect of various forms of heat treatment and plastic deformation (working) on the critical current density/external magnetic field relationship is indicated, as well as the effect of the state of the matrix on this relationship.

It is usually considered that a conductive toward the maintenance of superconductivity in high magnetic fields is a state of the material in which a system of well-dispersed superconducting and normal regions is realized when the field is applied, namely, either the so-called "mixed" state of A. A. Abrikosov [1] or a "Mendelsson sponge" [2]. Inhomogeneity of the material is required in order to realize either of these: in the first case in order to fix the superconducting filaments, and in the second in order to form regions with different critical fields.

Known hard superconductors acquire their high properties only after appropriate treatment, especially plastic deformation and the decomposition of the saturated solid solution (for materials based on a bcc lattice) or specific conditions of diffusion annealing (for materials based on metallic compounds).

It is considered that, in a high field, a system of superconducting formations (a kind of lattice) is maintained; although this only occupies a small volume, it is able to pass a considerable superconduction current.

Except for Bean's experiments [3] the study of hard superconductors has generally been carried out with materials which had been superconductors over their whole volume before the application of the magnetic field. However, in order to study the characteristics of the collapse of superconductivity in such materials we must study material initially constituting a dispersed mixture of phases.

Material and Experimental Method

In our own investigation we attempted to secure a metallic system in which, under certain conditions of heat treatment, we could obtain dispersed inclusions of a precipitating superconducting phase in a matrix normal at the test temperature. Earlier it was found that an alloy of copper containing up to 2 wt.% niobium could become superconducting at 4.2°K after prolonged annealing at 700 to 800°K. It is possible that the decoration of the boundaries and subboundaries of the copper with niobium may have created superconducting channels.

In this paper we present some data regarding the superconducting properties of an alloy of zirconium containing 4 wt.% Nb. According to the phase diagram this alloy constitutes a

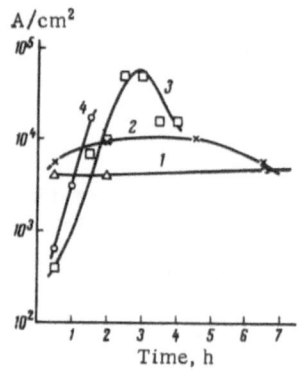

Fig. 1. Critical current density of a Zr–4%Nb alloy as a function of the time of decomposition of the supersaturated solid solution at 550°C. Degree of deformation, %: 1) 0; 2) 40; 3) 93; 4) 98.

solid solution based on a bcc lattice above 800°C [4]. By quenching, a supersaturated solid solution of niobium in hexagonal zirconium may be obtained.

Since the solubility of niobium in α-Zr is small, appropriate annealing may lead to the precipitation of the β-Nb phase, which has a T_K of over 4.2°K, while pure zirconium and zirconium containing traces of niobium in solid solution have a T_K of under 4.2°K. Thus as the supersaturated solid solution decomposes we obtain a system comprising a precipitating phase which is superconducting at 4.2°K within a matrix which is normal at this temperature.

The alloy was cast in an arc furnace, carefully annealed, quenched, and cold-worked with various degrees of reduction. Quenching and aging were carried out in a vacuum of $2 \cdot 10^{-6}$ mm Hg. In order to measure the superconducting properties, strip samples were used; the measurements took place in liquid helium at 4.2°K. The current was introduced smoothly by means of a semiconducting current amplifier. The transformation was recorded on a two-coordinate self-recording millivoltmeter.

The influence of a magnetic field was studied by tests in a superconducting solenoid. The samples were oriented parallel to the field.

Fig. 2. Microstructure of a Zr–4%Nb alloy (×600). a) After quenching from 1000°; b) after quenching and aging at 550° for 2 h.

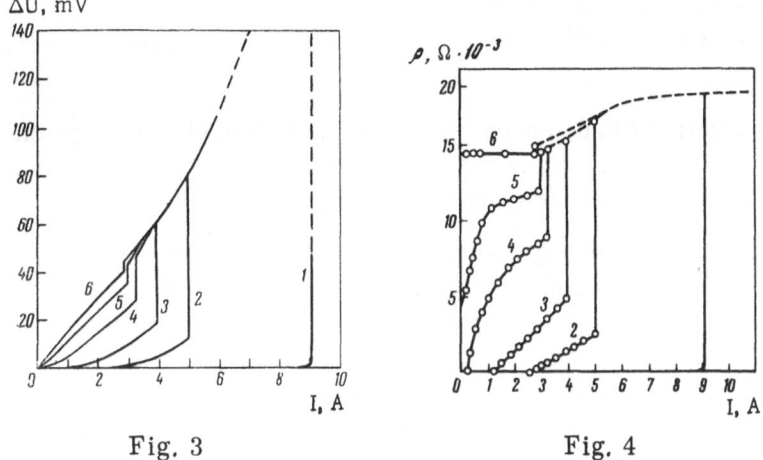

Fig. 3   Fig. 4

Fig. 3. Curves representing the collapse of superconductivity caused by a current in a longitudinal magnetic field. Alloy Zr–4%Nb, deformation 82%, aging at 550°. Field in kOe: 1) 0; 2) 1; 3) 2.2; 4) 5; 5) 10; 6) 20.

Fig. 4. Resistance as a function of the current introduced in a longitudinal magnetic field for a Zr–4%Nb alloy (notation as in Fig. 3).

Results and Discussion

All the samples subjected to quenching were nonsuperconducting at 4.2°K. The application of plastic deformation (cold-working) also produced no superconductivity at this temperature. Aging at temperatures above 500°C led to the appearance of superconductivity at the boiling point of helium [5].

Figure 1 shows the critical current as a function of the aging time at 550°C. A rise in critical current occurs over the first 2 to 3 h of heating. We see that the critical current density depends very much on the degree of cold work. For high degrees of deformation there is a maximum on the critical current-density curve. After three hours aging the critical current density of 98%-deformed samples reaches $7 \cdot 10^4$ A/cm^2.

On aging at 500°C analogous relationships are obtained. It should nerertheless be mentioned that samples not subjected to preliminary deformation gave no superconducting state even after 9 hours aging.

Micrographs confirm the decomposition of the supersaturated solid solution on aging. On quenching, there is a martensitic transformation of the β solid solution into a supersaturated α' solid solution, as indicated by the characteristic shear-transformation microstructure (Fig. 2a).

The aging responsible for the decomposition changes the microstructure revealed. We notice α crystals of zirconium decorated by the precipitating phase. The microsection darkens on etching (Fig. 2b).

Electron-microscope study showed [6] that aging was accompanied by complex processes of dislocation redistribution with the formation of a fairly dense dislocation lattice decorated by the precipitating phase; apparently the dense lattice is formed by the dispersed superconducting phase in the normal matrix.

The dimensions of the dislocation walls formed approximately equal 50 to 100 Å, while the density is of the order of 10^{10} cm^{-2}, which facilitates the passage of a current density of up to $7 \cdot 10^4$ A/cm^2.

The superconductivity is due solely to the precipitating phase, since the change in the composition of the matrix proceeds in the sense of a reduction in niobium concentration, which should lead to a fall in the T_K of the matrix.

Figure 3 shows the curves representing the disruption of superconductivity by a current in the presence of a magnetic field, as recorded on a two-coordinate automatically-recording millivoltmeter; the curves are specific properties of the material being studied. The horizontal axis gives the value of the current introduced and the vertical axis the voltage drop in the sample in mV. This kind of record enables us to trace the disruption of superconductivity from the appearance of a slight resistance to the complete restoration of the normal state.

We see that in the absence of a field the transformation is very sharp and takes place in one jump. However, only a slight field has to be applied in order to reduce the critical current substantially. The resistance appears gradually, while a jump is recorded after reaching a particular resistance value. With increasing field strength the resistance increases with the current more and more sharply; the "tail" associated with the smoothly-increasing resistance becomes longer and the magnitude of the sharp jump diminishes in accordance with this.

Finally, in fields of about 20 kOe we only see a very slight break on the transformation curve, indicating that only individual regions remain in the superconducting state.

Figure 4 shows the change in resistance as a function of the current introduced, as calculated from the earlier V/A curves. In small fields the rise in resistance proceeds linearly, the rate of rise increasing with the field. In high fields the rise in resistance follows curves of the saturation type. In fields above 10 kOe the resistance appears, as a rule, immediately, and changes little with increasing current, although a slight jump occurs. A similar picture of the disruption of superconductivity characterizes other forms of treatment.

Thus the disruption of superconductivity takes place over a wide range of current and field values; hence different parts of the sample have different relationships between the critical current and the magnetic field.

The rise in resistance on increasing the current load is associated both with the disruptive effect of the current itself and also with Joule heating in the parts of the samples which have been transformed into the normal state and the additional disruption of new parts of the superconducting chain.

Until the Joule heat is able to pass out into the bath of liquid helium there is a smooth rise in resistance; as soon as this heat outflow is disrupted (when more heat than the bath can carry away is evolved), there is a rapid heating of the sample and a sudden transformation into the normal state.

An analogous picture was observed by Bremer in tin films [7]. Using the pulse method, Bremer found that if the Joule heating were eliminated the transformation should be smooth right up to the point of reaching the normal state.

The revelation of an intermediate state in the alloy under consideration is facilitated by the fact that the amount of the superconducting phase is small and the area of contact with the liquid bath through which the heat is carried away relatively large.

The dependence of the critical current density on the applied magnetic field is shown for several samples in Fig. 5. We see that the shapes of the curves remain practically the same for samples receiving different heat treatment. The critical current for aging at 550°C is

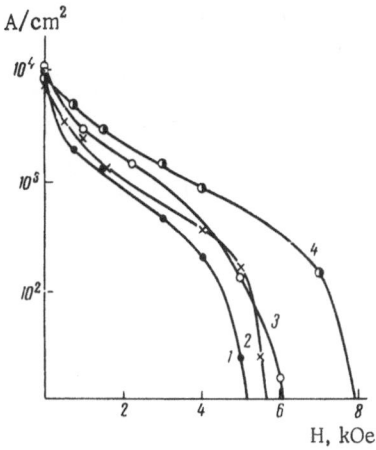

Fig. 5. Critical current density as a function of the longitudinal magnetic field (Zr–4%Nb). 1) Deformation 96%, aging at 500° for 70 h; 2) deformation 82%, aging at 500° for 95 h; 3) deformation 82%, aging at 500° for 4 h; 4) the same for 1 h.

slightly higher. However, for these samples T_K is also slightly higher (for aging at 550° T_K varies from 7 to 7.5°K and for aging at 500° from 5 to 7°K). It is premature to speak of any specific laws in this. Investigations into this aspect are continuing.

The considerable fall in critical current density may be explained, firstly, because the T_K of the alloy in the state in question is lower than in pure niobium (it varies up to 7.5°K); the reduction in T_K takes place under the influence of the normal matrix on the fairly thin interlayers of superconducting phase. This kind of influence on the part of the normal metal has been observed repeatedly [8, 9]. Secondly, in ordinary superconductors the magnetic field penetrates gradually and the dimensions of the superconducting regions vary accordingly.

In our own case, however, we have a material in which the superconductivity in the matrix has already been removed, the field penentrates immediately into a considerable proportion of the volume, and the critical current-density/field relation should correspond to the final part of the I/H curve for an ordinary hard superconductor.

In addition to this it is entirely possible that the precipitating phase fails to form a perfectly continuous lattice, and the zero resistance in cases of low fields occurs because the breaks in the circuit are overcome by virtue of the Josephson superconducting tunnel effect [10]. It is well known that the Josephson superconducting current is extremely sensitive to the magnetic field [11].

Usually the superconducting tunnel current overcomes dielectric gaps of the order of 10 Å; if, however, these discontinuities are filled with metal, then the extent of the tunnel gap may be greater, of the order of 10^{-5} cm [12].

In the presence of breaks in the superconducting circuit, we must allow for not only the effect of the magnetic field and current but also the heating of the sample by virtue of Joule heat evolution.

Thus we have found that highly-dispersed inclusions of a superconducting phase may form a multiply-connected superconducting system capable of passing a considerable current density. The critical density is estimated with respect to the cross-sectional area of the sample, whereas the superconducting part occupies only a small proportion of the volume, and the true current density in the superconducting circuit is at least an order higher.

Despite the relatively low values of T_K, the critical fields reach values of 5 to 8 kOe.

The disruption of the superconducting state takes place over a wide range of currents and magnetic field strengths, as is characteristic for hard superconductors.

LITERATURE CITED

1. A. A. Abrikosov, Zh. Éksp. Teor. Fiz., 32:1442 (1957).
2. K. Mendelsson, Proc. Roy. Soc. (London), A152:34 (1962).
3. B. C. Bean, M. V. Doyle, and A. G. Pincus, Phys. Rev. Lett., 9:94 (1962).
4. Yu. F. Bychkov, A. N. Rozanov, and D. M. Skorov, Metallurgy and Metallography of Pure Metals [in Russian], Atomizdat (1959).

5.   E. P. Romanov, L. V. Smirnov, V. D. Sadovskii, and N. V. Volkenshtein, Fiz. Met. Metalloved., 20:3 (1965).

6.   S. V. Sudareva, N. N. Buinov, V. A. Vozilkin, E. P. Romanov, and V. G. Rakin, Fiz. Met. Metalloved., 21 (3):388 (1966).

7.   J. W. Bremer and V. L. Newhouse, Phys. Rev., 116:309 (1959).

8.   P. H. Smith, S. Shapiro, J. L. Miles, and J. Nicol, Phys. Rev. Lett., 6:686 (1961).

9.   P. Hilsch and R. Hilsch, Naturwiss., 48:549 (1961).

10.   B. D. Josephson, Phys. Lett., 1:251 (1962).

11.   B. D. Josephson, Rev. Mod. Phys., 36 (1, Pt. 1):216 (1964).

12.   H. Meissner, Phys. Rev., 117:672 (1960).

ON RAISING THE ELECTRICAL RESISTANCE
OF ZIRCONIUM–NIOBIUM ALLOYS BY COOLING
TO CRITICAL TEMPERATURES

Yu. F. Bychkov, M. T. Zuev, V. A. Mal'tsev,
A. N. Rozanov, and I. S. Khukhareva

The effect of quenching temperature, degree of deformation, aging, and niobium content in the solid solution on the specific electrical resistance of Zr–15-25%Nb alloys at 4 to 300°K is considered. Alloys quenched from high temperatures increase their specific electrical resistance on cooling from room temperature to the temperature of the superconducting transformation, while after cold working or aging the temperature coefficient of resistance takes the positive value normal for metals. The reasons for this behavior of the alloys are discussed.

Alloys corresponding to a wide range of concentrations of the Zr–Nb system are particularly interesting as materials for making superconducting solenoids designed to produce a field of 70 to 100 kOe. These alloys have high values of the critical field: An alloy with 75% Nb has $H_{K2} = 80$ kOe at 4.2°K, while the maximum value of H_{K2} in the Zr–Nb system (120 kOe) occurs for a Zr–25%Nb alloy [1].

It is well known that one of the most important superconducting characteristics on which the possibility of using an alloy for the manufacture of solenoids depends, the critical current density, is a structure-sensitive property, i.e., it reacts to the existence of any kind of crystal-lattice imperfections or dispersed inclusions, particularly those of the nonsuperconducting kind or those with poor superconducting characteristics. Hence a detailed study of the structural characteristics of these alloys and their physical properties is of particular interest in order to understand the relation between the superconducting characteristics and structures of the alloys.

We studied some anomalous (for metallic systems) characteristics of the variations arising in the electrical resistance of Zr–Nb alloys on cooling to low temperatures, as far as the temperature corresponding to the transformation into the superconducting state. A measurement of the resistance at low temperatures is particularly interesting in view of the fact that the upper critical field of the alloys depends on its value [2].

We see from Fig. 1 that the electrical resistance of niobium-base alloys falls sharply on cooling to nitrogen temperatures, while the resistance of wire or strip made from zirconium alloys containing 15 to 20% Nb changes very little in the same temperature range (by 2 to 5%). These alloys have a small but positive temperature coefficient of resistance after deformation, as characteristic of metallic systems. Alloys with small quantities of niobium have a considerable positive temperature coefficient of resistance. The residual resistance in zirconium and niobium is a few percent of the resistance at room temperature.

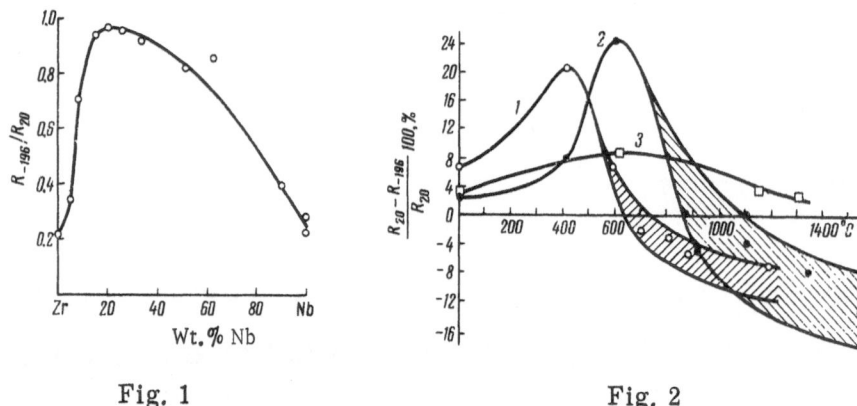

Fig. 1   Fig. 2

Fig. 1. Ratio R_{-196}/R_{20} for cold-worked Zr–Nb alloys.

Fig. 2. Relative variation in the electrical resistance of Zr–Nb alloys with quenching temperature. Nb content (%): 1) 15; 2) 20; 3) 25.

In view of the fact that the resistance of alloys containing 15 to 20% Nb varied very little with temperature, we thought it would be interesting to discover whether these alloys could have structural states giving them a zero or negative temperature coefficient of resistance.

We therefore took samples in the form of wires, strips, and rods, heated these in the β region at temperatures of 700 to 1300°, and then cooled them rapidly.

Two methods of heating were used: 1) In evacuated quartz bulbs, in which the wires were placed after wrapping in zirconium foil in order to prevent oxidation on water cooling; 2) in an apparatus of the UVR-2 type, in which the wires were heated in vacuum by passing a current and cooled by disconnecting the latter.

We see from Fig. 2, which shows the change in ρ (%) as a function of quench temperature, that after quenching from the region of the β phase the properties of alloys containing 15 to 20% Nb changed: On cooling to nitrogen temperatures their resistances rose, and on further cooling as far as the temperature of the superconducting transformation they remained almost constant (Fig. 3). Usually on cooling superconducting materials to the critical temperature their electrical resistance falls monotonically to a certain value, and then suddenly falls to zero. In the alloys under consideration the resistance rises by a few percent on cooling to 7 or 8°K and then falls sharply to zero. In contrast to cold-worked alloys of the same composition, the transformation into the superconducting state is spread over a range of width 0.5°K for the quenched alloys, not 0.1°K as in the cold-worked material (Fig. 4).

In the quenched state the alloy containing 15% Nb also has some other special features, in particular a very low elastic modulus (about 5000 kg/mm^2) and a high specific resistance (over 100 $\mu\Omega \cdot$ cm), exceeding that obtained by extrapolating the ρ/composition curve to 15% Nb.

The negative temperature coefficient of resistance in the alloy containing 15% Nb quenched from the β phase (from 1200°C) is retained even on heating to 200 or 300°, as shown earlier in [3].

Plastic deformation (working) of the alloy containing 15% Nb leads to a change in the sign of the temperature coefficient of resistance: It becomes positive and increases to a certain limiting value with increasing degree of deformation (Fig. 5).

For an alloy containing 20% Nb the temperature coefficient increases. Aging at 400 to 500° also leads to an increase in the temperature coefficient of resistance. It is important to note that on increasing the Nb content from 15 to 20% the temperature of quenching from the β

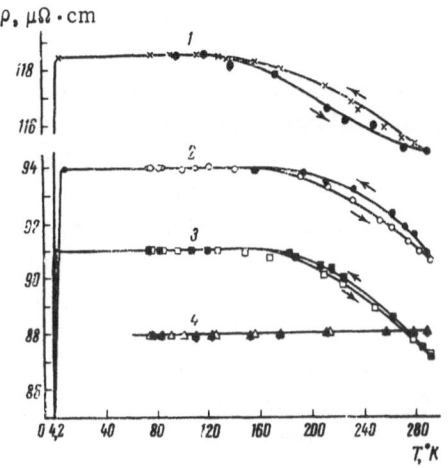

Fig. 3. Change in the specific electrical resistance of quenched Zr−Nb alloys on cooling. 1) 15% Nb, quenched from 850°; 2) 20% Nb, quenched from 1100°; 3) the same from 900°; 4) the same from 850°.

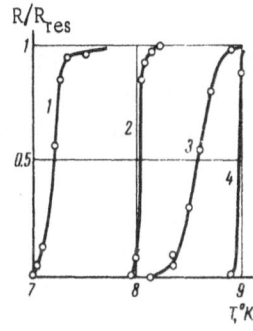

Fig. 4. Transformation of quenched and cold-worked Zr−Nb alloys into the superconducting state. 1) Alloy containing 15% Nb quenched from 850°; 2) the same, cold-worked; 3) alloy containing 20% Nb quenched from 1100°; 4) the same, cold-worked.

region for which the alloys begin taking a negative temperature coefficient of resistance increases, as in Fig. 2.

This combination of superconductivity with a negative temperature coefficient of resistance (anomalous for metals) and a high absolute resistance (on the borderline between metals and semiconductors) is not only found in quenched Zr−Nb alloys but also in a number of other quenched or cast alloys. It has been observed in particular in the following alloys: Ti−9.1−11.85 at.%Fe; Ti−7 at.%Mo; Hf−12.2 at.% Mo; Zr−8.6 at.%Mo [4], and also in Ti−Cr alloys.

V. N. Gridnev and V. N. Trefilov [5], discussing the reasons for an analogous anomalous rise in conductivity on heating Ti−Mo alloys, suggested that the metastable β phase might have a semiconducting nature, on the grounds that Ti−Mo alloys had a specific resistance of the order of 140 $\mu\Omega\cdot$cm and a negative temperature coefficient of resistance.

In the opinion of these authors, the width of the forbidden band for the metastable β phase was 10^{-2} to 10^{-3} eV, i.e., very narrow. The possibility that short-range ordering might occur was rejected on the grounds that a study of the background intensity on the x-ray diffraction pictures failed to support the existence of any marked short-range order in Ti−Mo alloys.

The reduction in conductivity on reducing the temperature may be due to the semiconducting nature of the β phase, as proposed by Gridnev and Trefilov [5], but may also be associated with transformations in the alloys taking place during heating and cooling. Since the electrical resistance of the alloys at liquid-nitrogen temperature was independent of changes in the rate of cooling, from very high (600 deg/min) to very low (1.4 deg/min), such transformations, if they occur, should take place very rapidly.

Repeated measurements of electrical resistance on successive heatings and coolings gave identical results, i.e., these transformations must be reversible.

The alloys under consideration contain the minimum proportion of niobium required to ensure that on quenching from the β phase the β solid solutions should be completely fixed; for lower Nb concentrations the ω and α phases separate on quenching [3].

In order to discover the peculiarities of the structural state of alloys containing 15 to 20% Nb quenched from the β region, we carried out some x-ray diffraction experiments. In view of the fact that deformation led to a change in the sign of the temperature coefficient, the x-ray photographs were obtained from large grains on the basis of the Laue (reflection) method, using

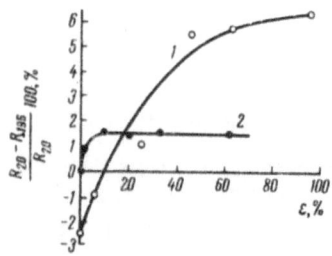

Fig. 5. Relative change in the electrical resistance of zirconium alloys containing 15% Nb (1) and 20% Nb (2) as a function of the degree of cold work.

an x-ray microbeam. The use of a sharp-focus tube with a copper anode enabled us to reduce the exposure to 2 or 3 h. The x-ray diffraction pictures (Fig. 6) show sharp reflections from the matrix (bcc lattice) and diffuse rings which may be regarded as a consequence of diffuse scattering from segregations formed on quenching. The diffuseness of the rings indicates that the segregations have the form of plane discs. The two-dimensional discs forming the rings by reflection must be regularly oriented relative to the original matrix: They are evidently arranged in planes parallel to the {001}, since on taking photographs along directions differing from the [001] the intensity of the diffuse rings is weaker. It was shown in [6] that the ω phase formed on quenching constituted a very defective structure and comprised thin plates parallel to the {110} planes of the β phase.

Regarding the composition of the segregations we may consider that these differ in no way from the ω phase, since on aging these alloys that is the phase which separates. The x-ray study showed that alloys containing 15 and 20% Nb did not constitute a pure β solid solution in the quenched state, but contained two-dimensional segregations regularly oriented in the matrix and connected coherently to the latter. Since the separation of the two-dimensional zones of ω phase takes place mainly by way of a martensitic transformation, the number of these may change on changing the temperature, even for low values of the latter.

An x-ray diffraction picture taken at liquid-nitrogen temperature differed slightly from one taken at room temperature from the same point of the sample: At liquid-nitrogen temperature all the sharp reflections were preserved, but the intensity of the diffuse rings fell slightly.

The change in the intensity of the diffuse rings at low temperatures indicates that the observed anomalies may be associated with the reverse $\beta \rightarrow \omega$ martensitic transformation, leading to the formation of a large number of two-dimensional segregations on cooling below room temperature and to their dissolution in the matrix on heating from liquid-helium to room temperature. For each temperature there must be a specific number and size of the segregations. The observed rise in the temperature coefficient of resistance on increasing the temperature of quenching from the β region is associated with the fact that at high temperatures the grains coarsen and the concentration of the vacancies formed on quenching becomes greater, as a result of which the conditions for the formation of two-dimensional segregations of the ω phase become more favorable: The grains are more supersaturated with vanancies and impurities at

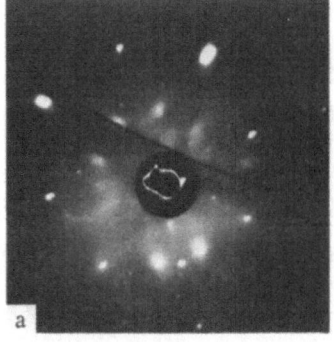
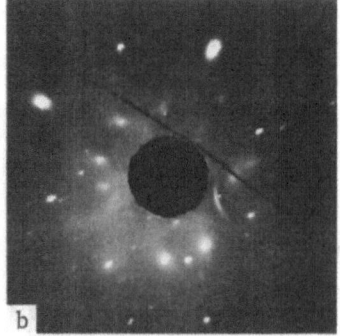

Fig. 6. X-ray pictures of quenched Zr–15%Nb taken at –100 and +23°C (a and b respectively).

the expense of a reduced concentration of these at the grain boundaries and a reduction in the length of the boundaries themselves. On raising the quench temperature from 900 to 1100° the electrical resistance at room temperature rose by 2 or 3%.

As shown by Guinier, the dimensions of the segregations formed on quenching usually depend on the homogenization temperature and increase as this rises [7]. It is also well known that the concentration of vacancies in quenched alloys containing 15% Nb greatly reduces the incubation period for the formation of the ω phase [8].

The rise in the temperature coefficient of resistance after cold-working is evidently associated with the formation of incoherent nuclei of the ω phase on deformation. Both cold work and aging promote the formation of ω-phase nuclei and suppress the stage of segregation formation (preprecipitation stage). On the formation of two-dimensional segregations coherently connected with the matrix, the resistance rises slightly, but on separation of the ω-phase nuclei the resistance tends to fall: After prolonged aging at 240°C the electrical resistance of the alloy containing 15% Nb rose by 10%, after aging at 300°C it remained almost unchanged, while aging at 350 to 450°C greatly reduced the electrical resistance. Evidently near 300°C the distortions introduced into the lattice by the segregations raise the resistance to the same extent that the particles of ω phase, by eliminating the stresses in the lattice, reduce it, so that on aging at 300°C the resistance remains constant.

The existence of a reverse martensitic transformation taking place on cooling a quenched alloy containing 15% Nb is confirmed by Fedorov, Kissil', and Zhomov's results based on measurements of internal friction at nitrogen temperatures. These authors observed a substantial rise in internal friction for temperatures below room temperatures; the change in internal friction was reversible [9].

LITERATURE CITED

1. T. G. Berlincourt and R. R. Hake, Phys. Rev. Lett., 9 (7):293 (1962).
2. A. A. Abrikosov, Zh. Éksp. Teor. Fiz., 32 (6):1442 (1957).
3. Yu. F. Bychkov and A. N. Rozanov, in: "Metallurgy and Metallography of Pure Metals" [in Russian], No. 1, Izd. MIFI (1959), p. 224.
4. R. R. Hake and D. H. Leslie, J. Appl. Phys., 34 (2):270 (1963).
5. V. N. Gridnev and V. N. Trefilov, in: "Physics of Metals and Metallography" [in Russian], Izd. AN Ukr. SSR, No. 14, p. 5 (1962).
6. B. A. Hatt and J. A. Roberts, Acta Metallurgica, 8 (8):575 (1960).
7. A. Guinier, Inhomogeneous Metallic Solutions [Russian translation], IL, Moscow (1964).
8. Nuclear Sci. Abstracts, US Atomic Energy Commission, 15 (1):91 (January 15, 1961).
9. G. B. Fedorov, A. E. Kissil', and F. I. Zhomov, Contributions to the Fourth All-Union Conf. on Relaxation Phenomena in Solids [in Russian], Tsent.-Chern. Izd., Voronezh (1965).

X-RAY DIFFRACTION STUDY OF THE FINE STRUCTURE OF A SUPERCONDUCTING Zr—25%Nb ALLOY

N. A. Sokolov, Yu. F. Bychkov, V. A. Mironenko, and A. A. Rusakov

The change in the dislocation density of a Zr—25%Nb alloy on plastic deformation and annealing is determined by subjecting the shape of the x-ray lines to harmonic analysis and applying the method of micro x-ray diffraction. A rise in critical current density may even take place on reducing the dislocation density but also on the presence of well-dispersed second-phase inclusions when these have poor superconducting properties.

One of the most important properties of superconducting alloys, the critical current density, depends sharply on the degree of plastic deformation (working) and the annealing conditions. This property is structure-sensitive. A determination of the mutual relationship between the critical current and the fine structure of the alloy is therefore a most important metallographical problem. However, hardly any work has been done in this connection.

We have therefore studied changes taking place in the structure of Zr—25%Nb after plastic deformation involving a 0 to 94% degree of reduction, after annealing the worked alloy, and after annealing a quenched sample.

For all the forms of treatment mentioned there was a rise in the critical current density. Thus after annealing the deformed wire at 400 to 570° the critical current density increases 30 times, reaching 10^5 A/cm^2 in fields of 30 kOe (Fig. 1). After plastic deformation (with high degrees of reduction), the critical current density for this alloy is about 10^4 A/cm^2 and as usual rises with increasing degree of deformation. On annealing the quenched alloy there is also a rise in critical current, although to a lesser extent than after plastic deformation [1].

Fine Structure of the Alloy After

Quenching and Deformation

As original sample for study we took Zr—25%Nb quenched from 900°C after holding for 1 h. X-ray diffraction showed that the alloy was a homogeneous β solid solution with a bcc lattice. The structure was nearly perfect, i.e., the sample contained no microstresses, and the dimensions of the regions of coherent scattering were quite large.

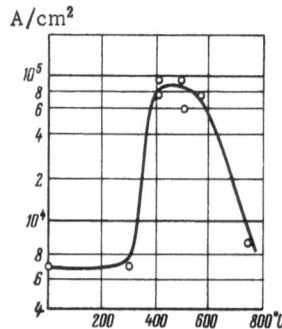

Fig. 1. Critical current density for a worked Zr—25%Nb alloy as a function of annealing temperature (duration 1 h).

Fig. 2. Microstructure of Zr–25%Nb samples. a) Quenched;
b) 10% deformed; c) 50% deformed.

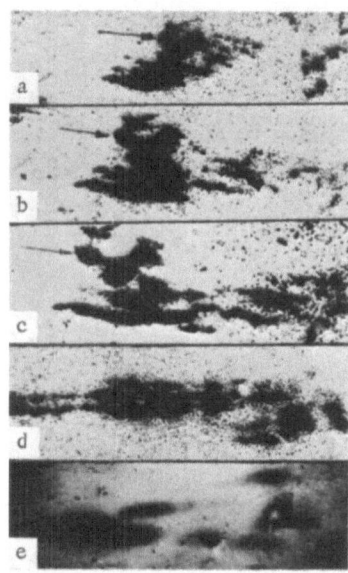

Fig. 3. Characteristic x-ray diffraction pictures obtained by the microdiffraction (MD) method. a) After quenching, distance from sample to film 5 mm; b) the same, 10 mm; c) the same, 15 mm; d) after quenching and 5% deformation; e) the same, 10% deformation.

This was supported by the sharp resolution of the K_α-doublet in the lines with large Bragg angles.

The samples were deformed (worked) by rolling at room temperature, the degree of plastic deformation being 5, 10, 15, 50, 75, and 94%. The microstructure of the samples is shown in Fig. 2. The change is fine structure was studied by the x-ray microdiffraction method (MD) with harmonic analysis of the diffraction-line shapes.

Owing to the severe blurring of the reflections for large degrees of deformation (reduction) the MD method was only able to reveal the substructural characteristics for the sample in the original state after 5, 10, and 15% reduction. Figure 3 shows the characteristic x-ray diffraction reflections from individual grains of the samples. The first photographs illustrate the method of determining the characteristics of the substructure, the diffraction reflections being recorded for several different sample-film distances. The method facilitates determination of the dimensions of the substructural constituents and their disorientation angles (more details of the MD method in [2]).

In the case of the quenched sample and also after 5% deformation, a substructure comprising two orders occurs: large blocks of the first order (about 120 μ) and smaller ones of the second order (10 to 20 μ). From the sizes of the mosaic blocks and the block-disorientation angles the dislocation density in the samples may be calculated from the formula

$$\rho_{bl} = \frac{\Delta\psi}{3bl}; \quad \rho_{bound} = \frac{\alpha}{bl}; \quad \rho_{tot} = \rho_{bl} + \rho_{bound}.$$

Here $\Delta\psi$ is the block disorientation angle, α is the disorientation angle between the blocks, b is the Burgers vector, and l is the mosaic block size.

Even slight plastic deformation causes substantial changes in the mosaic structure and dislocation density (Table 1).

In all the deformed samples the change in fine structure was also studied by harmonically analyzing the diffraction-line shapes. The interference maxima were recorded on a URS-50I diffractometer with discrete recording. The correction for instrumental broadening was introduced by the Stokes method [3], the standard being a specially-chosen fine-grained, recrystal-

TABLE 1. Fine-Structure Characteristics of the Samples

State of the sample	Second-order blocks		First-order blocks			$\rho \times 10^{11}$ 1/cm^2
	l, μ	$\Delta\psi$, min	l, μ	$\Delta\psi$, min	α, min	
Quenched	16.6	4.6	120	25	36	2.6
5% deformed	11.3	9.0	170	34	48	3.6
10% deformed.	—	—	120	58	100	8.0
15% deformed.	—	—	90	98	170	14.0

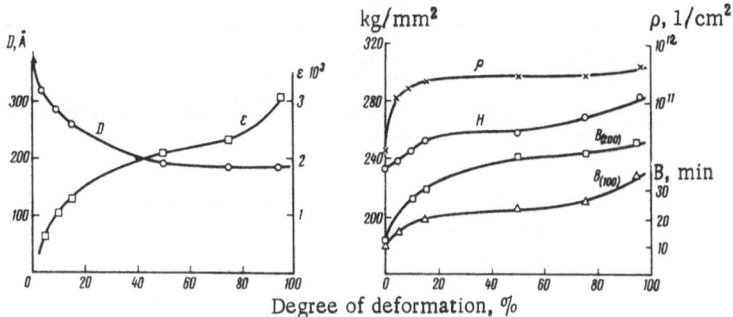

Fig. 4. Changes in the lattice microstresses ε, the size of the coherent scattering regions D, the dislocation density ρ, the microhardness H, and the width of the (200) and (110) diffraction lines of Zr–25%Nb in relation to the degree of plastic deformation.

lized molybdenum sample. The fine-particle-size and microstress effects were separated from one another by the Warren–Averbach method, using two orders of reflection from the (110) plane [4].

From the fine-structure characteristics obtained, the size of the regions of coherent scattering D, and the lattice microdistortions ε, we calculated the excess dislocation density in the samples, using the Williamson and Smallman formulas [5]. We see from Fig. 4 that the character of the changes in microhardness, dislocation density, and line width during the plastic deformation of the samples is approximately the same: a sharp rise on deformation up to 20%, and then a continuous but slower increase.

It should be noted that for all the samples except the original the (200) reflection was anomalously diffuse as compared with (110) and (211) lines. This is probably associated with the preferential arrangement of defects along specific crystal planes on deformation. The alloy samples had a strong texture after rolling, the (100) plane being parallel to the rolling plane, and the [100] direction parallel to the rolling direction. This kind of texture system is usually found in metals with a bcc lattice. This was shown to be the case in particular for Zr–50%Nb, which also had the structure of a β solid solution. In this connection it would be interesting to study the fine structure of samples by reference to the (200) and (400) interferences; however, the latter could not be recorded accurately enough owing to its low intensity and severe blurring.

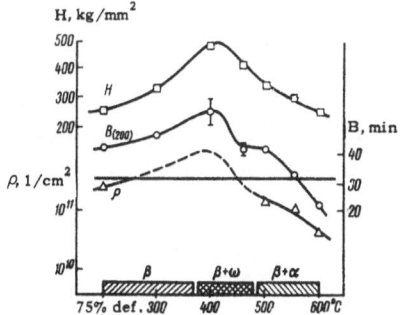

Fig. 5. Variation in the microhardness H, line width B(200), and dislocation density ρ for Zr–25%Nb samples as functions of aging temperature in the course of thermomechanical treatment.

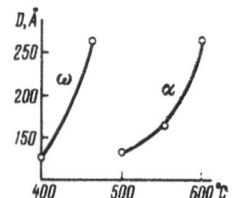

Fig. 6. Particle sizes
of the ω and α phases
in the samples as func-
tions of the aging tem-
perature.

Fine Structure of the Alloy

After Heat Treatment

Samples deformed by 75% were annealed at 300, 400, 460,
500, 550 and 600°C for 1 h.

It is well known that the metastable ω-phase precipitates in
an alloy of this composition at 400 to 460°, while at 500° or over the
α phase (almost pure Zr) separates out.

In the region corresponding to the precipitation of the ω-phase
from solution it was impossible to separate out maxima from the
β phase suitable for harmonic analysis, since this method requires
a precise knowledge of the behavior of the line near the background, and the lines of the ω phase
were close to the lines of the β phase being analyzed. We therefore had to analyze the variations in the width of the β-phase line; although more coarsely this procedure also offers a
possibility of estimating the fine structure in the β solid solution.

After annealing at 400° there is a considerable broadening of the line as compared with
the cold-worked state (Fig. 5). This is evidently associated with precipitation of finely-dispersed particles of the ω-phase, which distort the crystal lattice of the matrix.

By the method of harmonic analysis we were only able to determine the fine-structure
characteristics for samples annealed at 500, 550, and 600°. The method of calculation was the
same as before, using the (200) and (400) reflections. From the results the dislocation density
in the β-solid solution was found.

Figure 5 shows the variation in dislocation density from 400 to 460° as a broken line, by analogy with the variation in the width of the (200) line. The straight line on the graph indicates the
dislocation density of a 94%-deformed sample. It is an interesting fact that the dislocation density in samples annealed at 500° and above was smaller than that in the sample deformed by 94%.

Hence a rise in critical current density may even take place on reducing the dislocation
density in the matrix, or it may be associated with other factors, in the present case with the
presence of dispersed second-phase inclusions (ω or α) in the matrix of the β solid solution.

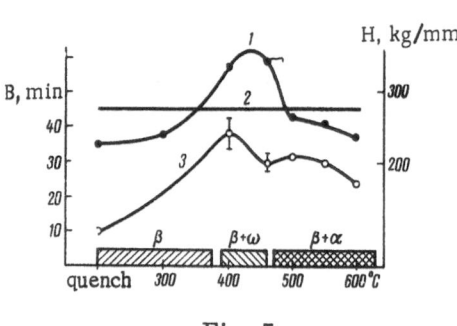   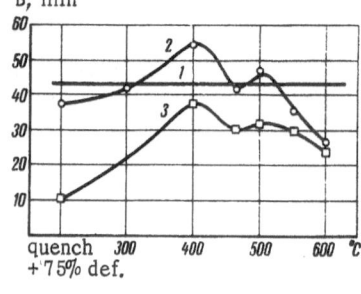

Fig. 7   Fig. 8

Fig. 7. Changes in the microhardness H and line width B(211) for samples
of Zr—25%Nb after various kinds of treatment: 1) Microhardness after
quenching and aging; 2) width of the (211) line after 94% deformation; 3)
the same after quenching and aging.

Fig. 8. Change in the line width B(211) for Zr—25%Nb alloys in various
original states as a function of aging temperature. 1) Alloy 94% deformed;
2) alloy quenched and 75% deformed; 3) alloy quenched.

Assuming the absence of microstresses in the precipitates of the ω and α phases, we may use the width of the diffraction line to estimate the size of the particles of the ω and α phases (Fig. 6). The particles have minimum dimensions after annealing at temperatures corresponding to the onset of precipitation of the new phase. For higher temperatures the size of the particles increases. Clearly the assumption that there are no microstresses present is imprecise under normal conditions. On allowing for the stresses (which are difficult to determine) the dimension being determined increases by several times.

In order to separate the influences of the ω-and α-phase precipitates and the preliminary deformation on the fine structure, we studied the change in structure resulting from the annealing of quenched samples. We found that on precipitation of a new phase the lattice of the matrix became severely distorted as compared with original state (Fig. 7). The line width had a maximum value on precipitation of the ω phase and diminished for higher-temperature annealing. In all cases the line widths of the annealed samples were smaller than those of the 94% cold-worked material, although after annealing at 400° the lines were similar in magnitude.

The change in the phase composition on annealing a quenched alloy is the same as in the annealing of cold-worked samples, but the decomposition of the solid solution occurs more slowly, so that the maximum microhardness is associated with higher temperatures.

Figure 8 compared the change in diffraction-line width on annealing, for alloys in different original states. We see that on annealing a quenched sample the line width is much smaller than on annealing a worked alloy. The maximum difference occurs in the initial state; on raising the annealing temperature the curves approach each other.

On the basis of the foregoing investigation our conclusions are as follows:

1. On plastically deforming (working or annealing quenched samples) there is an increase in the distortion of the crystal lattice; the effect is the greater for plastic deformation. The rise in critical current density observed in the case of the deformation or annealing of quenched samples matches the increase in dislocation density.

2. During the annealing of worked alloys at temperatures above 500° there is a reduction in dislocation density. At the same time the critical current density in these samples is 20 to 30 times greater than in severely-worked alloys.

Hence the rise in the critical current density does not only come from the increased dislocation density but depends on other factors also, in particular on the presence of highly-dispersed inclusions of the ω and α phases, which have poor superconducting properties.

LITERATURE CITED

1. Yu. F. Bychkov, I. N. Goncharov, and I. S. Khukhareva, Zh. Éksp. Teor. Fiz., 48 (3):818 (1965).
2. V. M. Kardonskii and I. P. Kushnir, Zavod. Lab., No. 6, p. 765 (1961).
3. A. Stokes, Proc. Phys. Soc., 61:382 (1948).
4. B. E. Warren, Progr. in Metal Phys., 8:147 (1959).
5. G. K. Williamson and R. Smallman, Phil. Mag., 1:34 (1956).

II. EFFECT OF THERMOMECHANICAL TREATMENT ON THE SUPERCONDUCTING PROPERTIES OF Nb—Zr, Nb—Ti, AND Nb—Hf BINARY ALLOYS

EFFECT OF HEAT TREATMENT ON THE TEMPERATURE CORRESPONDING TO THE TRANSFORMATION OF Nb—Zr ALLOYS INTO THE SUPERCONDUCTING STATE

V. V. Baron

The superconducting transformation temperature of Nb–Zr alloys is measured after cold working and various forms of heat treatment leading to changes in the phase composition of the alloys. The transformation temperature of the alloys changes after heat treatment; as in the case of single-phase alloys, it depends on the chemical composition of the phases precipitating.

The transformation temperature (T_K) of single-phase superconducting alloys changes with changes in their chemical composition. The character of the composition/T_K curves in the single-phase regions of various binary alloy systems depends chiefly on how the main physical characteristics determining the transition temperature in accordance with BKS theory change on alloying.

The highest transformation temperature in niobium–zirconium alloys occurs at a concentration of 25 at.% Zr, and in niobium–titanium alloys at 50 at.% Ti [1-4]. The alloys of these systems undergo decomposition of the high-temperature β solid solution over wide concentration and temperature ranges, these processes being due to the existence of polymorphism in zirconium and titanium. As shown in [5], the critical current of the alloys may be considerably raised as a result of these processes.

The influence of decomposition processes taking place during heat treatment on the transformation temperatures has been less studied.

We measured the transformation temperature of Nb–Zr alloys after cold working and various forms of heat treatment leading to a change in the phase composition of the alloys. The temperature at which the alloys passed into the superconducting state was measured by the method developed in the A. A. Baikov Institute of Metallurgy.* The results of the measurements of T_K were compared with the phase composition of the alloys determined by an x-ray method.

Figure 1, based on the results of our own measurements, shows the variation in the transformation temperature of Nb–Zr alloys on raising the Zr concentration to 80%. We see that the transformation temperature rises from 9.4°K for electron-beam-melted niobium to about 11.2°K for a concentration of 25 to 27 at.% Zr. Further increasing the zirconium content reduced T_K to 7.5°K in the alloy containing 80% Zr.

---

*See the article "Apparatus for Measuring the Temperature Corresponding to the Transformation of Metals and Alloys into the Superconducting State," this volume, p. 191.

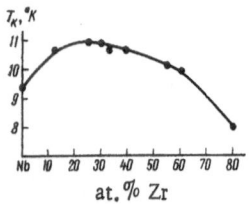

Fig. 1. Temperature of the transformation into the superconducting state for niobium—zirconium alloys.

In order to study the effect of heat treatment we took alloys containing 25 and 50 at.% Zr. The cold-worked samples were annealed at 500 to 900° for various periods in order to obtain different phase compositions of the alloys.

The original cold-worked alloy samples had the structure of the high-temperature β solid solution with a bcc lattice. The lattice parameter was 3.37 Å in the alloy with 25% Zr and 3.42 Å in the alloy with 50% Zr. On annealing these alloys, the decomposition of the β solid solution took place in accordance with the phase diagram, the velocity of this process being much greater than in cast alloys owing to the preliminary cold work, which accelerates the diffusion processes.

Thus after annealing at 900° for 15 min the alloy with 25% Zr showed particles of β_1 phase separating out from the solid solution and having a lattice constant of 3.32 Å. On increasing the holding period to 60 min the structure of the alloy comprises phases β_1 and β_2. The transformation temperature of the alloy with 25% Zr changes little as a result of these transformations (Fig. 2), since the chemical composition of the phases precipitating as a result of the decomposition at this temperature differs little from that of the β solid solution. This is because the annealing temperature is close to the upper boundary of the region of monotectoid decomposition of the alloy. A slight rise in T_K may probably be explained by a partial relief of the cold-work stresses.

Annealing at 800° reduces T_K from 11.2 to 10.7°K (holding time 120 min), since there is in this case a greater change in the chemical composition of the phases as compared with the original alloy.

A still greater fall in transformation temperature occurs after annealing at 600°. In this case no decomposition of the solid solution is revealed by the x-ray method, the transformation temperature evidently being more sensitive to the initial stages of decomposition. The greater fall in T_K at this temperature (from 11.2 to 10.5°K) may be explained by the greater difference in chemical composition between the precipitating phase and the original β solid solution, which has a higher transformation temperature.

It is interesting to follow the influence of a very long holding time (up to 100 h) at 700° on the transformation temperature of this alloy (see Fig. 2). With increasing holding time and hence an increasing amount of the second phase, which has the lower transformation temperature, the

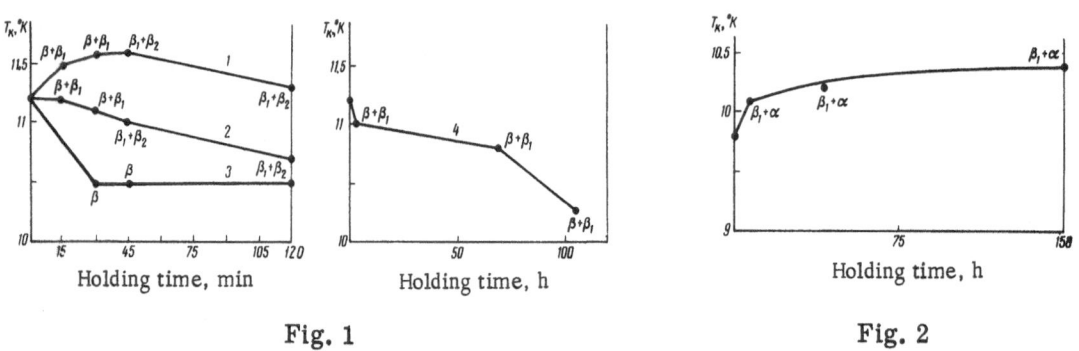

Fig. 1   Fig. 2

Fig. 2. Transformation temperature and phase composition of a cold-worked Nb—25%Zr alloy as a function of annealing period at 900° (1), 800° (2), 600° (3) and 700° (4).

Fig. 3. Phase composition and transformation temperature of a cold-worked Nb—50%Zr alloy as a function of annealing time at 500°.

T_K of the alloy falls still more, to 10.3°K. As a result of the heat treatment of the alloy containing 50% Zr (which in the single-phase state has a lower transformation temperature than the alloy with 25% Zr), there is a rise in T_K. After annealing at 860° for one hour, T_K rises to 11°K; this is due to the precipitation of the β_1 phase, which contains a greater proportion of niobium.

Annealing the Nb–50%Zr alloy at 500° also leads to a rise in T_K; after 150 h this becomes 10.4°K (Fig. 3). In this case, after precipitation of the hexagonal α phase, the β phase contains over 55% Nb.

Conclusions

1. The temperature corresponding to the transformation of Nb–Zr alloys into the superconducting state after various kinds of heat treatment depends on the phase composition of the material.

2. As in the case of single-phase alloys, the T_K of alloys having a two-phase structure is determined by the chemical composition, but in this case the variation in transformation temperature depends on the amount and composition of the precipitating phases.

LITERATURE CITED

1. J. K. Hulm and R. D. Blaugher, Phys. Rev., 123 (5):1569 (1961).
2. E. M. Savitskii and V. V. Baron, Izv. Akad. Nauk SSSR, Metallurgiya i Gornoe Delo, No. 5, p. 4 (1963).
3. H. Richter, P. Wincierz, K. Anderko, and U. Zwicker, J. Less-Common Metals, No. 4, p. 252 (1962).
4. V. V. Baron, M. I. Bychkova, and E. M. Savitskii, in: "Metallography and Metallophysics of Superconductors" [in Russian], Izd. "Nauka" (1965), p. 53.
5. V. V. Baron, in: "Metallography and Metallophysics of Superconductors" [in Russian], Izd. "Nauka" (1965), p. 29.

INDIVIDUAL EFFECTS OF PLASTIC DEFORMATION
AND THE DECOMPOSITION OF THE SOLID SOLUTION
ON THE CRITICAL CURRENT OF SUPERCONDUCTING
ALLOYS OF THE Nb—Ti AND Nb—Zr SYSTEMS

V. V. Baron and M. I. Bychkova

The separate effects of plastic deformation and the decomposition of the solid solution as a result of annealing Nb–Zr and Nb–Ti alloys on the superconducting parameters is studied. Both these processes increase the critical current of the alloys. The existence of decomposition processes in the alloys enables the critical current to be increased by over an order for a particular degree of decomposition. A high degree of preliminary deformation accelerates decomposition and in this case a high critical current may be achieved after shorter annealing periods than in the case of recrystallized material.

The fundamental parameters of superconductors (temperature of passing into the superconducting state, critical magnetic field, and critical current) depend on the chemical composition and structure of the alloys.

The critical current is the parameter most sensitive to a change in the structure of the alloys. Thus on changing the structure of Nb–Zr alloys by deformation (working) and annealing in such a way as to produce a decomposition of the solid solution, the critical current may be increased more than 30 times as compared with that of material not so treated.* The observed rise in critical current after working and heat treatment is associated with the consequent large number of dislocations, point defects, and second-phase precipitates, and also the redistribution of interstitial impurities at certain stages of the decomposition of the solid solution.

The existence of these physical and chemical inhomogeneities in the alloys promotes the flow of high currents in the alloys promotes the flow of high currents in superconductors of the second kind in the mixed state (above H_{K_1}) and offers extensive prospects of their practical application.

The question as to what form and size of inhomogeneity produces the greatest rise in critical current in superconductors of the second kind is a complicated problem; it may very well be solved by correlating the characteristics of the superconductors with an examination of their fine structure and a determination of various physical and mechanical properties at helium temperatures.

A study of the structure and properties of the same alloys subjected either to cold work or to heat treatment should provide us with certain information regarding the influence of these

---

*See an article by V. V. Baron "Superconducting Niobium–Zirconium Alloys and the Effect of Heat Treatment on Their Properties" in the collection "Metallography and Metallophysics of Superconductors" (Izd. "Nauka," 1965).

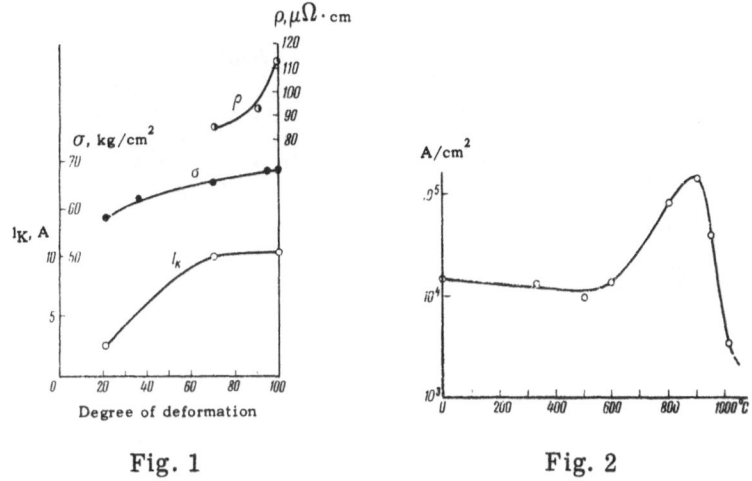

Fig. 1   Fig. 2

Fig. 1. Critical current, tensile strength, and electrical resistance
of a Nb—75%Ti alloy as a function of the degree of deformation.

Fig. 2. Critical current density of a Nb—50%Zr alloy as a function
of annealing temperature.

processes on the critical current of superconducting materials. We undertook such an investi-
ation in the case of Nb—Ti and Nb—Zr alloys (75% Ti, 33 and 50% Zr).

The samples were prepared in the form of wires from cast homogenized alloys having
(according to x-ray diffraction) the structure of the β solid solution. In order to eliminate the
influence of surface phenomena the cold working of the samples was carried out to varying ex-
tents in such a way that, after a recrystallizing anneal and further deformation to a final wire
diameter of 0.25 mm, the total degree of deformation should vary between 20 and 99.9%. As a
consequence of this, after cold working (from 20 to 95% deformation) the critical current of
niobium alloys containing 33% and 50% Zr increased from 3 to 30 and 2 to 18 A respectively.
In the alloy containing 75% Ti the critical current rose from 2.5 to 11 A on changing the degree
of deformation from 20 to 99.9%. The ultimate tensile stress (tensile strength), electrical resistance, and temperature of passing into the superconducting state were measured for the
same samples.

Figure 1 shows the influence of the degree of deformation on the critical current, tensile
strength, and electrical resistance of a Nb-Ti alloy. The critical current was measured from
the appearance of a voltage drop in a transverse external magnetic field (26.4 kOe) created by
a superconducting solenoid. The sharpest rise in critical current as a result of cold working
occurred between 20 and 70%. On further raising the degree of deformation the rise in critical
current became slower. The critical current density after 99.95% deformation was 2.10^4 A/cm^2.
As a result of cold working there was also a considerable rise in the tensile strength and electrical resistance.

As indicated in the foregoing, after heat treatment of the cold-worked Nb—Zr alloys, leading to decomposition of the solid solution, there was a still greater rise in the critical current.

Figure 2 shows the dependence of the critical current of an Nb—50%Zr alloy on annealing
temperature after holding for 1 h. Cold worked Nb—33-50%Zr wire was subjected to recrystallization annealing. In order to select the annealing conditions the recrystallization temperature
of the samples was determined. This equalled 1100 and 1050° for the alloys containing 33 and
50% Zr respectively. After annealing and quenching from these temperatures, the samples retained the single-phase structure of the β solid solution (according to x-ray diffraction). The
critical current fell to between 2.5 and 2 A at the same time. Subsequent annealing of these
alloys at 700 and 500° raised the critical current again (Fig. 3). The annealing conditions

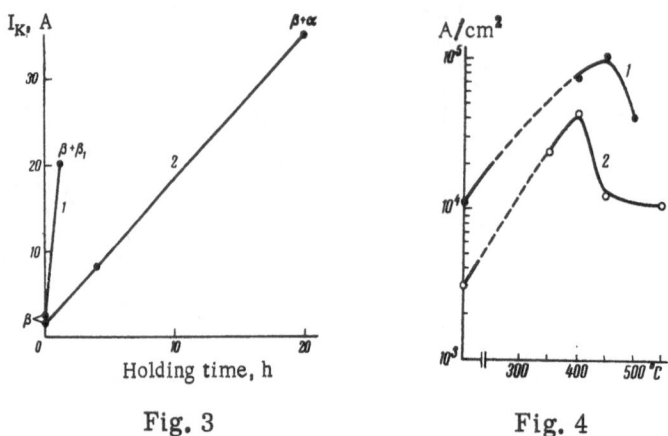

Fig. 3

Fig. 4

Fig. 3. Phase composition and critical current of Nb—Zr alloys as functions
of annealing time and temperature: 1) 700° (33% Zr); 2) 500° (50% Zr).

Fig. 4. Critical current density of worked (1) and recrystallized (2) Nb—75%Ti
alloys as functions of annealing temperature.

chosen enable us to judge the effect of the structure of the second phase precipitating during
the decomposition of the solid solution on the critical current. Higher currents were obtained
as a result of the separation of the hexagonal α phase from the bcc β phase. In order to achieve
this a longer period of annealing at 500° was required (because of the lower rate of diffusion in
the alloys). It is interesting to note that on using a heat-treated wire for the winding of a superconducting solenoid consisting structurally of the ($\beta + \alpha$) phases instead of a wire with the
$\beta + \beta_1$ structure (two bcc phases) the degree of "degeneracy" diminished.

The effects of heat treatment on the characteristics of the Nb-Ti alloys were studied in
more detail. Wire samples of this alloy were recrystallized at 800° (the recrystallization temperature was determined metallographically) and subjected to further annealing in order to
decompose the solid solution (350 to 550° for 1 h). Figure 4 (curve 2) shows the effect of annealing temperature on the critical current density of such a sample; with increasing temperature this rises, reaching a maximum at 400°, then begins falling. For comparison the same
figure (curve 1) shows the effect of annealing temperature on the cold-worked wire (99.95% deformation). Although the measurements showed that the highest values of j_K were achieved as
a result of a combination of cold work and annealing, there was also a considerable rise in this
characteristic for previously recrystallized samples, by an order of magnitude at 400°. The
lower critical current of the recrystallized wire samples after annealing may be explained by
the lower velocity of the decomposition processes in this case as compared with the previously-
worked wire. This is also indicated by the greater changes in electrical resistance, tensile
strength, and superconducting transformation temperature of the previously-worked samples
(Fig. 5). In this case there is a greater fall in electrical resistance and a greater rise in the
transformation temperature and tensile strength. On x-ray diffraction analysis of the alloys
annealed at 400 to 550°, after a holding period of 1 h the hexagonal α phase was always found
to have been formed.

In order to increase the degree of decomposition of the solid solution the annealing time
of the samples was increased (up to 100 h) as the optimum temperature (450°).

As a result of these experiments (Fig. 6) it was found that an increase in the holding
time led to a rise in the current of the recrystallized samples. The critical current density
increases particularly rapidly in the range 1 to 10 h, but after 100 h only equals 10^5 A/cm²; at

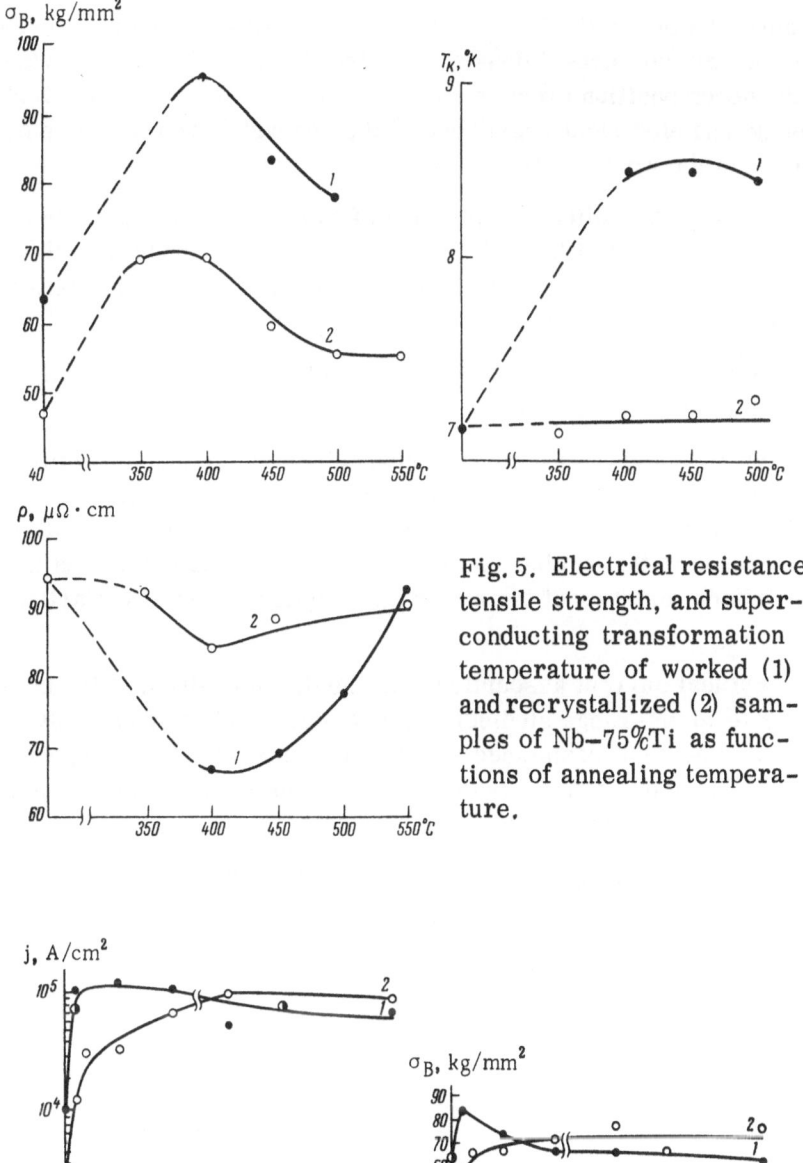

Fig. 5. Electrical resistance, tensile strength, and superconducting transformation temperature of worked (1) and recrystallized (2) samples of Nb–75%Ti as functions of annealing temperature.

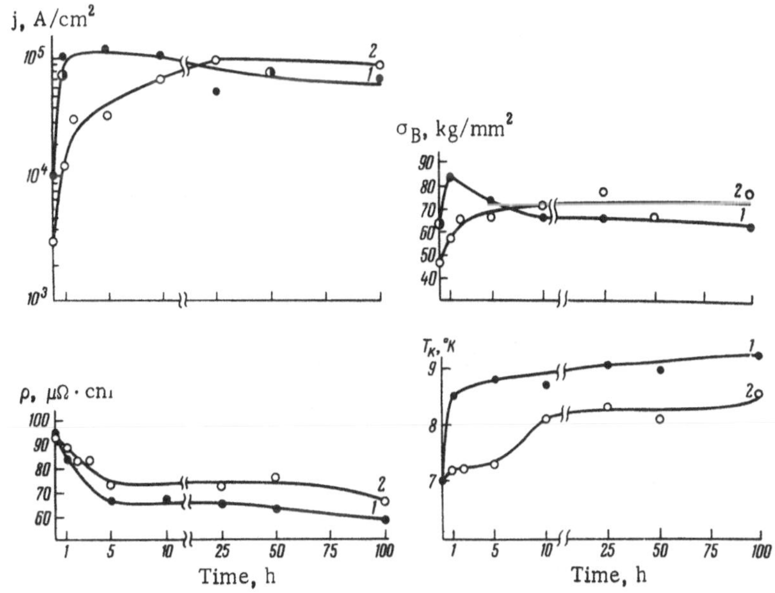

Fig. 6. Superconducting transformation temperature, critical current density, tensile strength, and electrical resistance of worked (1) and recrystallized (2) samples of Nb–75%Ti as functions of the period of annealing at 450°.

the same time there is a fall in the j_K of cold-worked alloys after holding periods of over 10 h. The reduction in the critical current density and also the tensile strength evidently indicates a continuation of the decomposition process and the onset of softening in the worked samples. The tensile strength and electrical resistance of the recrystallized wire are higher after such annealing that those of the worked wire.

Analysis of the separate effects of plastic deformation and the decomposition of the solid solution on the critical current of the Nb–Zr and Nb–Ti superconducting alloys showed that both physical and chemical inhomogeneities arising during these processes tended to raise the characteristics of the superconductors.

At a certain stage in the decomposition of the solid solutions in the alloys under consideration, the critical current increases by more than an order.

A preliminary high degree of deformation accelerates the decomposition processes, and in this case a considerable critical current may be achieved as a result of comparatively brief annealing periods (as compared with the recrystallized materials). The softening of the alloys and clearly, an increase in the number and size of second-phase inclusions (above the optimal value) worsen the characteristics of the alloys and counteract the effect arising from plastic deformation.

The rise in critical current associated with the decomposition of the solid solution in recrystallized alloys (over the range studied) was not accompanied by any softening processes. The observed fall in electrical resistance and rise in transformation temperature on annealing may be explained as being due to the enrichment of the solid solution with niobium as a result of the precipitation of a zirconium-rich phase.

The results obtained may help in choosing methods of raising the critical current of certain superconductors and using these more efficiently.

EFFECT OF HEAT TREATMENT ON THE
SUPERCONDUCTING PROPERTIES OF Nb–Ti ALLOYS

M. I. Bychkova, V. V. Baron, and E. M. Savitskii

The effects of annealing time and temperature on the structure and superconducting characteristics (critical current and superconducting transformation temperature) of Nb–Ti alloys are studied individually. Heat treatment raises the critical current density of cold-worked alloys. The maximum critical current density is achieved in a Nb–75%Ti alloy (10^5 A/cm^2 in a field of 26 kOe) after annealing at 450° (for 1 h in final annealing or 10 h after intermediate annealing). The transformation temperature rises to 8.6°K in the annealed alloys. The effect of heat treatment on the superconducting characteristics is associated with the decomposition of the original β solid solution of the cold-worked Nb–Ti alloys.

Alloys of the niobium–titanium system are of interest in connection with the windings of superconducting solenoids. The critical fields of these alloys are very high compared with alloys of other systems incorporating transition metals forming continuous solid solutions [1, 2].

In previous investigations [3] we found that a failing of these alloys was the low critical current obtained in cast, worked samples. Published data relating to individual alloys of this system tend to agree with our own results [4-5].

At the present time it has been established that, in the case of superconducting alloys in which decomposition processes take place on reducing the temperature, the critical current may be considerable raised by appropriate heat treatment.

In the niobium–titanium system a continuous series of solid solutions is formed between niobium and β-titanium over the whole range of concentrations above the temperature of the polymorphic transformation in titanium. On the titanium side there is a region corresponding to the decomposition of the bcc β solid solution into the hexagonal α phase and a cubic phase richer in niobium than the original. The boundary of the solid solution cannot be established exactly below 500 to 600° owing to the very slow rate of diffusion. Some research workers have observed a compound NbTi in the region of 50 at.% Ti [6-7]. The boundary for the formation of the β solid solution in quenched alloys of the Nb–Ti system is given in a paper by Bagaryatskii et al. [Dokl. Akad. Nauk SSSR, 122:4 (1958)].

The aim of the present investigation was to study the possibility of obtaining alloys of the Nb–Ti system with high critical currents by changing the phase composition as a result of heat treatment in the range of concentrations corresponding to high critical fields. We also studied the structure of the alloys and the laws governing the changes in critical current after various forms of heat treatment. For these purposes we took niobium alloys containing 50 to 80 at.% Ti.

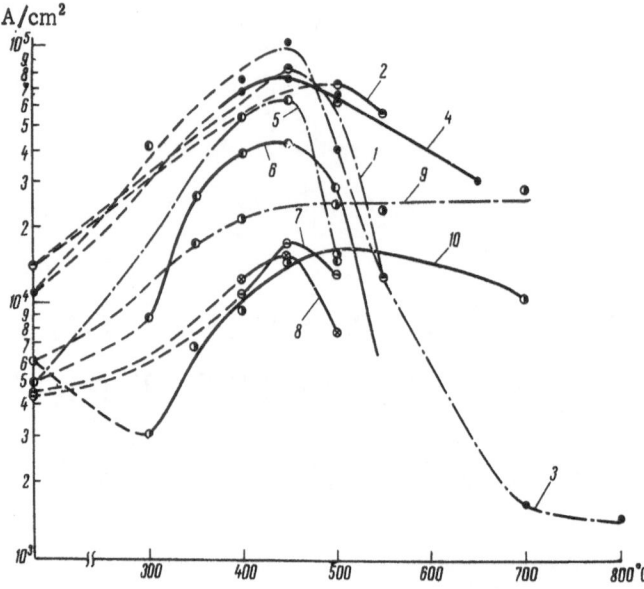

Fig. 1. Dependence of the critical current density of
Nb–Ti alloys on annealing temperature (annealing time
1 h). Titanium content in %: 1) 80 (final anneal); 2)
the same (intermediate anneal); 3) 75 (final anneal);
4) the same (intermediate anneal); 5) 70 (final anneal);
6) the same (intermediate anneal); 7) 65 (intermediate
anneal); 8) 60 (intermediate anneal); 9) 50 (final anneal); 10) 50 (intermediate anneal.

Method

The alloys were melted in an arc furnace with a nonexpendable electrode in an atmosphere
of purified helium (0.5 atm). The cast samples, 6 × 6 mm in section were cold rolled in a
grooved mill to dimensions of 1.3 × 1.3 mm, after which some were drawn through a Pobedit
(tungsten alloy) die into wire 0.25 mm in diameter.

Either the 1.3 × 1.3 mm samples or the wire 0.25 mm in diameter were subjected to heat
treatment. After intermediate annealing the undrawn samples were drawn to a diameter of 0.25
mm so that the cold work should restore the stresses partly removed by annealing.

The just-drawn samples were sealed into evacuated quartz ampoules with loose niobium
and titanium chips and annealed in a muffle furnace. After annealing they were cooled in water.

The tensile strength, specific electrical resistance, critical current (in fields from 0 to
26 kOe), and transformation temperature of the alloys were studied on wire samples.

The hardness (cast samples) was measured on an apparatus of the Brinell press type with
a load of 60 kg; the tensile strength was determined on a miniature tensile tester to an accuracy of 0.1 kg in load measurement. The electrical resistance was determined by a potentio-
metric method (accuracy of measuring the voltage drop in the wire 0.001 mV). Debye photographs of wire samples were taken in filtered copper radiation. The transformation temperature (T_K) was measured with a gas thermometer at temperatures above 4.2°K by reference to
the change in the magnetic properties of the sample. The thermometer was calibrated from
reference points. The accuracy of measuring T_K was of the order of 0.2°. The length of the
measured samples was 7 mm and the diameter 4 mm.

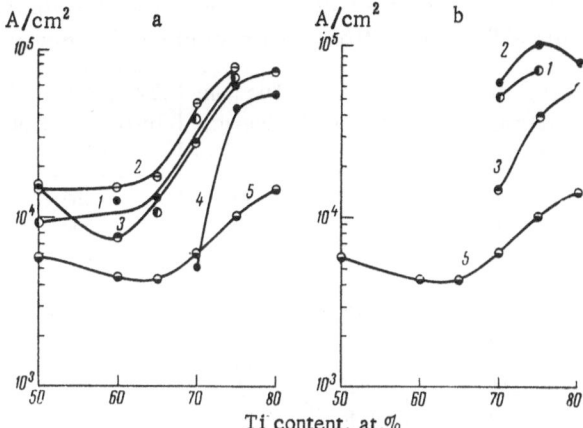

Fig. 2. Dependence of the critical current on the composition of the alloys. a) Intermediate anneal; b) final anneal. Annealing temperature, °C: 1) 400; 2) 450; 3) 500; 4) 550; 5) worked alloys.

The critical current of the alloys (wire 120 mm long) was measured in an apparatus situated in a transverse magnetic field created by an electromagnet with superconducting rings made from an Nb–Zr alloy. The transformation of the samples into the normal state was detected by reference to the appearance of a voltage drop equal to 50 mV.

Results of the Investigations

By annealing the samples between 300 and 550° a change in the phase and structural state of the alloys associated with the decomposition of the β solid solution was achieved. Annealing at 700 to 900° (above the boundary of the two-phase region) was also tried in order to establish the effect of recrystallization processes on the critical current of the alloys.

According to x-ray data, all the original cold-worked samples were single-phase and had a bcc lattice. The microstructure of the alloys was also single-phase.

The critical currents of the alloys under consideration largely depend on the annealing temperature and also on the composition of the heat-treated alloy (Figs. 1 and 2). For alloys with 60 to 75 at.% of titanium the optimum properties for the critical current are obtained after annealing at 400 to 450°. A reduction in critical current density occurs on raising the temperature to 500°. In the case of finally-annealed alloys the reduction in critical current density with increasing annealing temperature is sharper than in that of alloys subjected to intermediate

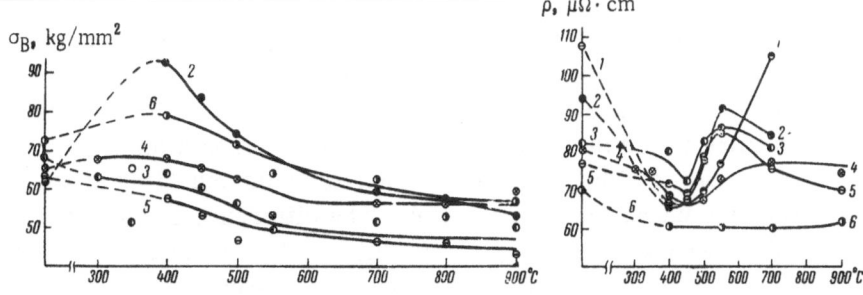

Fig. 3. Dependence of the tensile strength and specific electrical resistance of Nb–Ti alloys on the final annealing temperature. Titanium content, %; 1) 80; 2) 75; 3) 70; 4) 65; 5) 60; 6) 50.

TABLE 1. Temperature of Transformation into the Superconducting State of Niobium—Titanium Alloys after Various Forms of Annealing

Titanium content, at.%	Temperature of final* annealing (1 h), °C	T_K, °K	Titanium content, at.%	Temperature of final* annealing (1 h), °C	T_K, °K
50	worked	9.7	70	450	8.6
50	400	9.7	70	500	8.5
50	550	9.7	70	550	8.2
50	700	9.7	70	700	8.0
60	worked	9.6	75	worked	7.0
60	350	9.6	75	400	8.5
60	400	9.6	75	450	8.5
60	450	9.6	75	500	8.5
60	700	9.6	75	700	7.0
65	worked	9.1	80	worked	6.5
65	450	9.1	80	400	8.0
65	750	9.1	80	450	8.5
70	worked	8.0	80	700	6.5
70	400	8.6			

*Annealed at a final diameter of 0.25 mm

annealing. In an alloy with 75% Ti at 700° (final anneal) the critical current density is an order higher than in the original worked alloy; in an alloy of the same composition with intermediate annealing (700°) the critical current density is roughly equal to the original. For alloys containing a higher proportion of niobium, for example, those with 50% Ti, better results wer obtained after annealing at 500 to 700° ($j_K = 2.7 \cdot 10^4$ A/cm^2, final annealing). Owing to its brittleness after heat treatment, the alloy containing 80% Ti was not tested for all annealing temperatures; after annealing at 500° its characteristics were slightly higher than those of the alloy with 75% Ti.

In all cases involving the same alloys and the same conditions of heat treatment, the critical currents were higher after final annealing, as the degree of preliminary heat treatment was greater.

The highest critical current density was obtained for an alloy containing 75% Ti annealed at 450° ($1.02 \cdot 10^5$ A/cm^2).

The tensile strength of alloys with 60 to 70% Ti changes little on f i n a l annealing (Fig. 3a); there is a slight reduction in tensile strength with increasing annealing temperature; the alloy with 75% Ti shows a slight maximum of tensile strength at 400 to 450°, and there is a slight maximum in the case of the alloy with 50% Ti at 400°. Alloys with 75, 65, and 50% Ti subjected to i n t e r m e d i a t e annealing have a tensile-strength maximum at 450°; the values of tensile strength are here greater than in the case of final annealing.

All alloys except that with 50% Ti have a minimum of the specific electrical resistance (Fig. 3b) between 400 and 450°; in the alloy with 50% Ti there is a slight fall in specific electrical resistance starting from 400 and remaining up to 900°. The curves of electrical resistance for intermediate annealing differ in character from those obtained after final annealing. The electrical resistance of alloys with 75 and 70% Ti increases on annealing at 400 to 450°, and in the other cases it remains unchanged on annealing. The difference in these properties during preliminary and final annealing is probably due to the fact that the phase precipitating as a result of the decomposition of the solid solution serves as a source of additional stresses on deformation after annealing.

The transformation temperature of the alloys increases on annealing in the range of decomposition (350 to 550°). On annealing above the decomposition range the transformation tem-

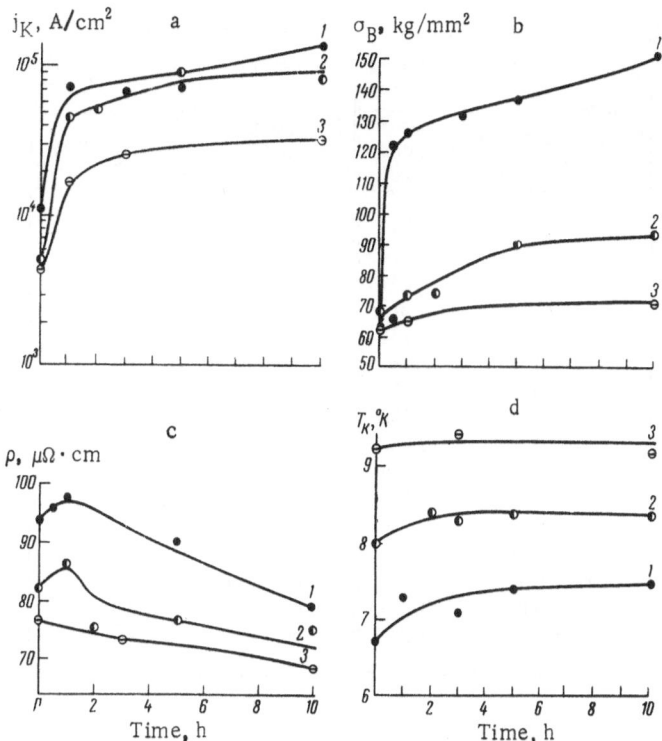

Fig. 4. Dependence of the critical current density, tensile strength, specific electrical resistance, and superconducting-transformation temperature of Nb–Ti alloys on the annealing time at 450°. Titanium content, %: 1) 75; 2) 70; 3) 65.

perature remains constant (Table 1). The greatest rise in transformation temperature as compared with the original occurs for the alloy with 80% Ti (from 6.5 to 8.5°K). The transformation temperature of the alloys containing less than 65% Ti remains unaltered after the annealing.

Increasing the annealing period to 10 h at 450° leads to a further increase in the critical current density of the alloys with 75, 70 and 65% Ti (Fig. 4a). In the alloy containing 70% Ti the current densities for periods of 5 and 10 h approach the critical currents of the alloy containing 75% Ti. The alloy containing 65% Ti lags on that containing 75% Ti as regards the rate of growth of critical current density, and after a 10-h anneal the rupture strength increases as compared with that of the original worked material.

The tensile strength also increases for all three alloys on increasing the annealing period (Fig. 4b). The greatest strength occurs for the alloy with 75% Ti. Both the strength and critical current of this alloy increase most in the first 30 to 60 min of annealing. On further increasing the holding period the curves flatten out.

The electrical resistance of these alloys falls slightly under the corresponding conditions. In alloys with 75 and 70% Ti the fall is preceded by a slight maximum (Fig. 4c).

The superconducting-transformation temperature of the lowest-temperature alloy among those considered (75% Ti) increases more rapidly that in the case of the alloy with 70% Ti. In the alloy with 65% Ti the transformation temperature remains unaltered after annealing for 10 h (Fig. 4d).

X-ray structural analysis showed that in the alloy with 80% Ti the second phase appeared after annealing for 1 h at 450 and 550°. The lattice of this phase was hexagonal. In the alloy

TABLE 2. Lattice Parameters of Niobium–Titanium Alloys After
Heat Treatment

Ti content, at.%	Temperature of intermediate annealing* (1 h), °C	Lattice parameter, kxu	Ti content, at.%	Temperature of intermediate annealing* (1 h), °C	Lattice parameter, kxu
0	worked	3.30	70	450 (10 h)	3.282
10	»	3.29	70	450	3.284
25	»	3.284	75	worked	3.28
50	»	3.27	75	450	3.26
50	450	3.264	75	500	3.262
60	450	3.28	75	550	3.28
65	worked	3.269	75	450 (2 h)	3.27
65	400	3.269	75	450 (3 h)	3.261
65	450	3.267	75	450 (5 h)	3.265
65	500	3.265	75	450 (10 h)	3.26
65	550	3.268	75	450 (50 h)	3.267
65	450	3.266	75	450 (100 h)	3.27
65	450 (10 h)	3.266	80	worked	3.28
65	450 (50 h)	3.26	80	400	3.262
70	worked	3.285	80	500	3.266
70	450	3.284	80	550	3.265

*Annealed at an intermediate diameter of 1.3 mm, samples then drawn to a diameter
of 0.25 mm.

with 75% Ti the second phase with the hexagonal lattice was observed at 450° after a 5-h holding
period, the lines being very weak.

In the remaining samples no second phase was observed by x-rays after heat treatment.

Texture was preserved in the alloys after annealing at 300 to 550°. At 700° and over the
alloys recrystallized.

In the alloys with 75 and 80% Ti the lattice parameter of the β solid solution falls to
values corresponding to the lattice parameter of a solid solution with a small titanium content.
Increasing the holding time at 450° to 10 h produces no change in the lattice parameter within
the limits of experimental accuracy. In alloys containing 65, 60 and 50% Ti the lattice parameter remains unchanged after annealing.

The results of an x-ray structural analysis of the alloys are presented in Table 2.

These investigations showed that as a result of the decomposition of the β solid solution
the critical currents of the alloys over the whole range of concentrations (50 to 80 at.% Ti)
could be appreciably raised. The corresponding fall in the electrical resistance and lattice
parameter and the rise in the tensile strength of certain alloys indicate the decomposition of
the β solid solution and the precipitation of a second phase. This process is accompanied by
improverishment of the base with respect to titanium (solid solution with a bcc lattice), which
leads to a reduction in the lattice distortions. The composition of the solid solution moves in
the direction of concentrations with lower critical currents than the original solid solution (results of [3]). Stresses of the second kind arising in the alloy as a result of deformation (working) are also reduced by annealing. Nevertheless, the critical current of the alloys rises at
certain stages of decomposition; we may suppose that this is because of the appearance of the
second phase.

On decomposition of the β solid solution, according to the phase diagram of this system,
a hexagonal α phase precipitates in alloys of the range of concentrations considered. In addition to this, at temperatures below 450° there may be precipitation of an intermediate ω phase.
In the alloy with 80% Ti the α phase is observed after 1 h at 450° and in the alloy with 75% Ti

after 5 h. The metastable ω phase is not revealed by x-ray diffraction. However, judging from the character of the curves representing the various properties, we cannot exclude the possibility that at temperatures of under 450° and for short holding periods (1 h) the ω phase may be formed in alloys with 75% of titanium or less. It is well known that the hexagonal α phase is superconducting with a transformation temperature of 0.37°K. On alloying with niobium the transformation temperature of the α phase increases, but never exceeds 1°K. The superconducting temperature of the ω phase in titanium alloys has not been studied; however, by analogy with zirconium alloys we may suppose that in the present case also it is not very high. Thus the rise in the critical current is associated with precipitation of a phase which has a transformation temperature much lower than that of the base of the alloy. On the basis of Andersen's mechanism [Phys. Rev. Letters, 9:309 (1962) for the passage of current through a superconductor of the second kind in the mixed state, the precipitating phase may fix the lines of magnetic flux. The force producing this fixing acts in opposition to the Lorentz forces arising when a current passes through a conductor situated in a perpendicular magnetic field. It was shown in a number of papers that a precipitating-particle size suitable for fixing was of the order of 10^{-5} cm. Thanks to their fairly low superconducting-transformation temperature and their precipitation in finely-dispersed form, both the α and the ω phase may fix the lines of magnetic flux and promote a rise in the critical current.

Judging from the properties investigated, the decomposition process takes place at a considerable velocity in the alloy with 80% Ti (after 1 h at 450° the hexagonal phase appears). With increasing niobium content the rate of decomposition becomes rapidly lower, as we should expect on the basis of the phase diagram.

It is interesting to note an anomaly in the properties of the alloy with 50% Ti (an increase in the tensile strength, and particularly critical current, after annealing). In view of the fact that certain authors have observed the compound NbTi in alloys with 50% Ti [6, 7], it is possible that the improvement in the properties as a result of annealing may be due to the initial stages in the formation of this compound; the critical current is most sensitive to this kind of phase change in the alloy.

Recrystallization-annealing at temperatures above the region of decomposition leads to a fall in the critical currents of the alloy by almost an order compared with the original worked material. If the recrystallizing anneal takes place at the intermediate diameter, the subsequent 97% deformation is sufficient to restore the properties to their original level.

Higher critical characteristics are obtained on annealing at the final diameter for an hour instead of annealing at the intermediate diameter because of the higher rates of decomposition in the former case. The accelerated decomposition is explained by the greater degree of deformation (99.97%) associated with the final annealing of the material rather than intermediate annealing, for which the degree of deformation is 94%. The reduction in the rate of decomposition for intermediate annealing is confirmed by the lower superconducting-transformation temperatures of alloys with 70 and 75% Ti as compared with those of alloys subjected to final annealing (see Table 1).

In order to increase the degree of decomposition on intermediate annealing in alloys with 75, 70, and 65% Ti the holding time at 450° was increased. In this way the critical current was raised considerably and so was the tensile strength, while the specific electrical resistance diminished. In alloys with 75% Ti the x-ray diffraction pictures showed the lines of a phase with a hexagonal lattice after a 5-h holding period.

Conclusions

As result of the present investigation it has been established that the critical current density of Nb—Ti alloys in the range of concentrations corresponding to high critical fields may be raised to $1.2 \cdot 10^5$ A/cm^2 or over by heat treatment designed to change the phase composition.

These results serve as a basis for selecting the compositions of Nb—Ti alloys and the methods of treating these.

In alloys of this range of concentrations the superconducting-transformation temperature increases as a result of the appearance of a bcc β phase richer in niobium than the original β-phase during the decomposition process.

Analysis of the critical current density/annealing temperature and critical current density/concentration curves (the latter for various annealing conditions), together with data relating to structure and electrical and mechanical properties, shows that the critical currents in Nb—Ti alloys increase on decomposition of the solid solution and the precipitation of a second phase (α or ω) from the latter. The composition inhomogeneities thus obtained in the alloy raise the value of the critical current, despite the fact that the stresses in the original bcc lattice diminish when the solution is impoverished with respect to titanium.

In character the curves relating tensile strength to the conditions of treatment are often very similar to the critical-current curves.

As a result of recrystallization of the material, the critical currents are an order of magnitude lower than in the original cold-worked Nb—Ti alloys.

LITERATURE CITED

1. T. G. Berlincourt and R. R. Hake, Phys. Rev., 131 (1):140 (1963).
2. C. K. Jones, J. K. Hulm, and B. S. Chandrasekhar, Rev. Mod. Phys., 36:74 (1964).
3. V. V. Baron, M. I. Bychkova, and E. M. Savitskii, in: Metallography and Physics of Superconductors [in Russian], Izd. "Nauka" (1965), p. 53.
4. R. R. Hake, T. G. Berlincourt, and D. H. Leslie, Superconductors, Proceedings of Technical Sessions (February 18, 1962), pp. 53-59.
5. T. G. Berlincourt, Symposium on Low-Temperature Physics, London (Sept., 1962), Vol. 3, p. 249.
6. V. P. Elyutin, M. L. Bernshtein, and Yu. A. Pavlov, Dokl. Akad. Nauk SSSR, 104:546 (1955).
7. N. V. Grum-Grzhimailo, Izv. Akad. Nauk SSSR, Otdel. Tekh. Nauk., No. 7, p. 24 (1957).

EFFECT OF THERMOMECHANICAL TREATMENT
ON THE SUPERCONDUCTING PROPERTIES
OF SOME Nb–Ti ALLOYS

Ya. N. Kunakov, E. V. Kachur, V. Ya. Pakhomov,
and D. I. Lainer

The effect of final heat treatment on the properties of certain Nb–Ti alloys is studied. The optimum final annealing temperature increasing the value of I_K lies between 450 and 500° and between 350 and 400°C for the alloys containing 30 and 50% Ti respectively. The effect of intermediate annealing on the critical current is also considered. It is shown that I_K rises from 7 to 20 A in an Nb–50%Ti alloy on carrying out intermediate heat treatment.

It was shown earlier in [1] and [2] that the superconducting-transformation temperature (T_K) of Nb–Ti alloys was about 10°K, and the critical magnetic field (H_K) of alloys with 4.3 to 4.5 valence electrons per atom reached 150 kG. Thus, whereas in superconducting solenoids with fields up to 50 or 60 kG it is better to use Nb–25%Zr and Nb–33%Zr alloys, which have a critical current density (j_K) higher than that of Nb–Ti alloys under analogous conditions, only Nb–Ti alloys may be used in order to obtain fields of the order of 100 kG.

A recent paper by Coffey, Hulm, et al. [3] mentioned the creation of a 100-kG solenoid in which the inner sections were wound from Nb–Ti alloys. It was also pointed out that the critical current of these alloys was less inclined toward degeneracy than that of Nb–Zr alloys.

However, the critical current density of cold-worked Nb–Ti alloys in fields of 50 to 60 kG is much lower than that of Nb–Zr alloys, reaching 3.10^4 and 1.10^5 A/cm^2 respectively. It would thus be desirable to increase the critical current density of Nb–Ti alloys by thermomechanical treatment. A study of the effect of thermomechanical treatment on the critical currents of Nb–Ti alloys is of particular interest, since according to the phase diagram [4] alloys with a low Ti content (up to 50 at.%) are single-phased, and in this way we should be able to obtain additional experimental data enabling us to extend our understanding of the nature of the changes in superconducting properties arising from thermomechanical treatment.

In order to study the effect of thermomechanical treatment on the superconducting properties of Nb–Ti alloys we selected alloys with 30, 50, and 70 at.% Ti. The Nb–30%Ti alloy lies in the region of the single-phase solid solution and undergoes no phase transformations on heat treatment. The Nb–70%Ti alloy may pass into a two-phase state after heat treatment at an appropriate temperature. Thus the alloys were chosen in such a way as to lie distinctly in either the one- or two-phase regions.

The original material for the alloys was thrice-remelted electron-beam-refined niobium and iodide-type titanium. The original bars were melted in an electric furnace with a tungsten electrode, then cold-forged to bars 11 × 11 mm and cold-drawn to their final dimensions. The

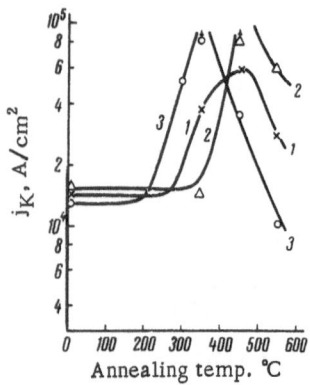

j_K, A/cm^2

Annealing temp. °C

Fig. 1. Dependence of the critical current density on the temperature of final annealing for cold-worked alloys. Titanium content, at.%: 1) 30; 2) 50; 3) 70.

samples were vacuum-annealed at a residual pressure of about 10^{-5} mm Hg for one hour. Before annealing the sample surface was cleaned mechanically. The critical current density was measured in a transverse magnetic field of 30 kG at 4.2°K. The values of j_K were estimated by reference to the first collapse of the superconducting state, i.e., without allowing for conditioning of the sample.

The effect of heat-treating the cold-worked samples at a final diameter of 0.25 mm on the properties of the various alloys was studied; so was the effect of one or several intermediate heat treatments on the j_K of Nb–50%Ti.

The effect of final annealing (samples 0.25 mm in diameter) on j_K was studied for all the alloys under consideration. The samples underwent 99.9% deformation in the cold state before annealing.

We see from Fig. 1 that in the cold-worked state the j_K of all the alloys was almost exactly the same. Heat treatment raises the critical current density independently of whether the alloys are one- or two-phased. The arrows in Fig. 1 indicate that the superconducting state was not disrupted for the particular current densities in question. The optimum annealing temperature for Nb–30%Ti and Nb–50% Ti alloys is 450° and for Nb–70%Ti is 350 to 400°.

As regards the nature of the rise in j_K in the superconducting alloys on heat treatment, there are no very clear views at the present time. Some authors [4] consider that the rise in j_K is due to phase transformations taking place on annealing (which may occur in such two-phased alloys as Nb–Zr); others associate these phenomena not with phase transformations but with a change in the structure of the alloy at the initial stage of recrystallization.

Even allowing for the possible influence of these factors in general on the change in j_K associated with the annealing process, it is hard to explain the observed results by this phenomenon, since heat treatment not only raises the j_K of alloys in which decomposition of the solid solution may take place in the course of annealing (Nb–70%Ti), but also that of alloys in which no phase transformations are to be expected at the annealing temperatures under consideration (Nb–30%Ti). The temperature of the onset of recrystallization in these alloys is about 800° for the specified degree of deformation. Probably in general the change in j_K during heat treatment is associated not only with possible phase transformations but also with a redistribution of defects in the alloy structure.

The effect of intermediate heat treatment on the critical current density of Nb–50% Ti was studied for samples 0.6 mm in diameter, which were later cold-drawn to 0.25 mm. The degree of cold deformation after annealing was 82.5%.

We see from Fig. 2 that intermediate annealing raises j_K in comparison with that of samples not so treated, even though the latter may have received a much greater degree of deformation (99.9%). It is also interesting to note that the temperature of optimum heat treatment in the present case (450°) coincides with the optimum temperature of final heat treatment. In addition to this, the introduction of intermediate annealing leads to a considerable improvement in technological properties.

The effect of several intermediate anneals on the j_K of Nb–50%Ti was also studied; annealing was carried out at 650° (after drawing the wire to 0.9 mm) and 560° (after drawing it to 0.55 mm).

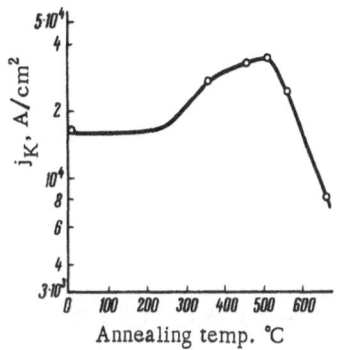

Fig. 2. Critical current density of an Nb–50%Ti alloy as a function of the temperature of intermediate annealing.

Measurements of the critical current density showed that this type of heat treatment also improved the properties of samples 0.25 mm in diameter in comparison with those in the cold-worked state. The critical current in the samples rose from 7 to 20 A, although the sample subjected to intermediate heat treatment had undergone much less deformation (70 to 80%).

Conclusions

1. We have studied the effect of final heat treatment on the properties of Nb–30%Ti, Nb–50%Ti, and Nb–70%Ti. The application of optimum heat treatment greatly increases the critical current density as compared with that of samples worked in the cold state. The optimum annealing temperature is 450 to 500° for the first two alloys and 350 to 400° for the Nb–70%Ti.

2. A rise in j_K as a result of heat treatment is observed both in two-phase and single-phase alloys; in general this effect is associated not only with possible phase transformations but also with a redistribution of defects in the structure.

3. The use of intermediate annealing raises the j_K in comparison with that of cold-worked samples and improves the technological properties of the alloys.

LITERATURE CITED

1. J. Hulm and R. Blaugher, Phys. Rev., 123:1959 (1961).
2. T. G. Berlincourt, R. R. Hake, and D. H. Leslie, Phys. Rev. Lett., No. 6, p. 671 (1961).
3. H. T. Coffey, J. K. Hulm, W. T. Reynolds, D. K. Kox, and R. E. Spian, J. Appl. Phys., 36 (1):128 (1965).
4. V. D. Borodich, A. P. Golub', A. K. Kombarov, M. G. Kremlev, N. K. Moroz, B. M. Samoilov, and V. Ya. Fil'kin, Zh. Éksp. Teor. Fiz., 44:1 (1963).

EFFECT OF VARIOUS FACTORS ON THE SUPERCONDUCTING PROPERTIES OF Nb—Ti ALLOYS

Ya. N. Kunakov, E. V. Kachur, and V. Ya. Pakhomov

The effect of heat treatment and the addition of up to 0.4 wt.% C on the superconducting properties of Nb alloys containing 30 to 70 at.% Ti is studied. By final heat treatment the critical current of the alloys may by substantially increased. After homogenization at 1500°C the critical current in worked samples falls sharply. However, after annealing the samples at their final diameter the effect of homogenization is leveled out and the critical current in the homogenized samples becomes the same as in the nonhomogenized material. On introducing carbon into the alloys the critical current rises sharply. The current rises particularly so after heat-treating the samples at their final dimensions.

It was shown earlier [1, 2] that the niobium—titanium system had critical fields of about 140 kOe. However, the critical current of these alloys is lower than that of the widely-used niobium—zirconium alloys. It is therefore of great interest to study the various factors leading to a rise in the critical current density of Nb—Ti alloys.

It was shown earlier* that final heat treatment had a considerable effect on the critical current in Nb—30%Ti, Nb—50%Ti, and Nb—70%Ti. In the present article we shall present data relating to the effects of a homogenizing anneal, the addition of carbon, and heat treatment at the final diameter in combination with intermediate annealing, on the critical current density of niobium alloys containing 46 to 70 at.% Ti.

Homogenization of the Alloys. It is well known that liquations (segregations) develop in alloys on solidification. Such inhomogeneities are also retained in the wire after cold drawing, and this will clearly have an effect on the superconducting properties of the latter. The intracrystalline liquation may be removed by a high-temperature homogenizing anneal; this will reduce various kinds of stresses and defects and will also plainly affect the superconducting properties. The influence of the cast state and the homogenizing anneal on the superconducting properties will clearly depend on the nature and composition of the alloys.

It was noted in [3] that for niobium alloys containing 28 and 33% Zr the annealing of the cast material for five days at 1250° led to a reduction of the critical current density in the wire samples (0.25 mm) as compared with that of the unannealed samples.

In the present investigation we studied the influence of a homogenizing anneal on the critical current density of niobium alloys containing 46 and 65% Ti (composition in at.%). We also partially studied the behavior of Nb—39%Ti—8%Zr, Nb—75%Zr and the alloy known as SS-2.

The alloys were melted in a cellular arc furnace with a tungsten electrode in an atmosphere of purified helium and subjected to a homogenizing anneal at 1500°C in a vacuum furnace of the TVV-4 type. The ingots weighed 40 g. After a homogenizing anneal the ingots were cold-

---

*See the preceding article: Effect of Thermomechanical Treatment on the Superconducting Properties of Some Niobium—Titanium Alloys.

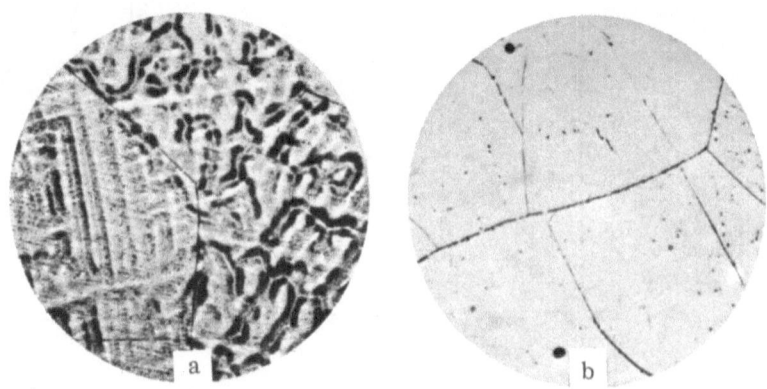

Fig. 1. Microstructure of Nb–46%Ti (× 200). a) In the cast state; b)
after a homogenizing anneal for 5 h at 1500°.

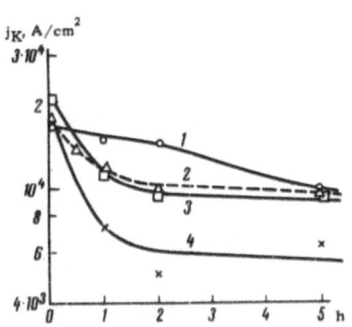

Fig. 2. Critical current density for worked samples as a
function of annealing time at
1500°. 1) Nb–46%Ti; 2) alloy
SS-2; 3) Nb–39%Ti–8%Zr; 4)
Nb–75%Zr.

-forged to bars 11 × 11 mm in cross section, and from these
0.25-mm wires were cold-drawn. All the samples had the
same degree of deformation in the cold state. Clearly in
this way the samples differed from each other simply in respect of the structure after the homogenizing anneal.

Figure 1 shows the microstructure of Nb–46%Ti before and after a homogenizing anneal. We see that in the
latter case the intracrystalline liquation has been removed
and the material has a structure of large polyhedral grains.

We measured the critical current density in the resultant samples in a transverse magnetic field of about 16 kOe.
Figure 2 shows that the critical current density in Nb–46%
Ti and also the other alloys falls considerably with annealing time.

We studied the effect of final annealing on homogenized
and unhomogenized samples of Nb–65%Ti. The results presented in Fig. 3 show that annealing at the final size (0.25

mm) levels the influence of homogenization. The critical current in the homogenized samples
in this case became even a little higher than in the unhomogenized samples.

The relationship obtained may be interpreted on the basis of the theory of Anderson [4].
It was shown earlier [5, 6, 7] that a "hard" superconductor in the mixed state was unstable with
respect to the superconducting transport current flowing in a direction perpendicular to the
magnetic field. Gorter [5] associated this instability with the appearance of Lorentz forces
acting on quantized magnetic lines (fluxoids) in "hard" superconductors during the flow of the
current. The appearance of Lorentz forces should lead to the motion of the fluxoids and to the
breakdown of the superconducting state. This kind of picture should appear in a homogeneous
"hard" superconductor. In the presence of inhomogeneities the latter prevent the motion of the
fluxoids, which form an aggregate at the inhomogeneities (form a bunch) and lie in a potential
well. In this case the Lorentz forces arising as a result of the flow of a transverse current
can only move the fluxoids if the current exceeds a certain value.

In the unhomogenized state, in the presence of various inhomogeneities due to intracrystalline liquation, the inhomogeneities prevent the motion of the fluxoids. Hence the critical

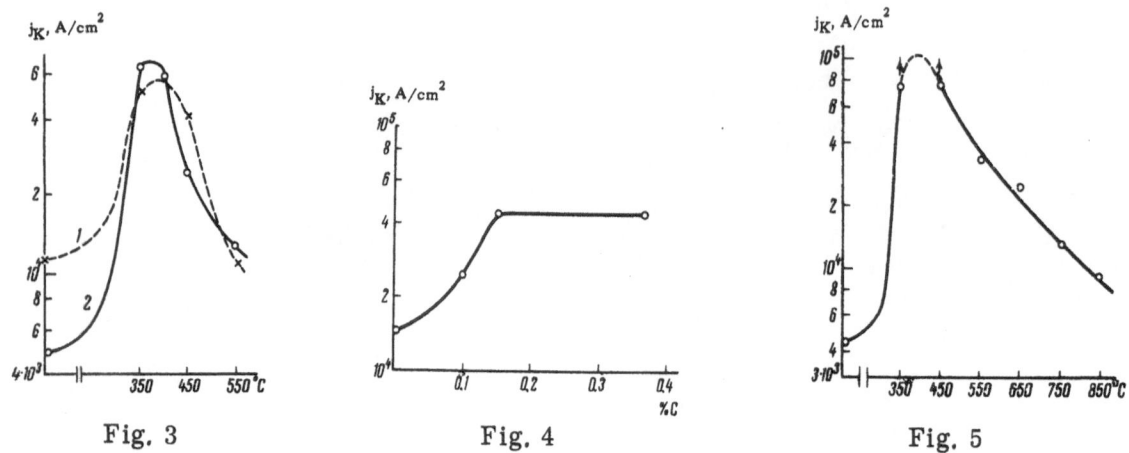

Fig. 3   Fig. 4   Fig. 5

Fig. 3. Critical current density as a function of the final-annealing temperature for un-homogenized (1) and homogenized (2) samples of Nb–65%Ti.

Fig. 4. Critical current density of cold-worked Nb–50%Ti as a function of carbon content.

Fig. 5. Critical current density of Nb–50%Ti samples containing 0.15% C, as a function of final annealing temperature.

current measured in a transverse magnetic field is higher for unhomogenized samples than for completely homogenized samples.

After heat treatment at the final diameter, new inhomogeneities develop in the samples; these are associated with decomposition or predecomposition in the solid solution. Hence after heat treatment at the final diameter the influence of the cast structure weakens, and processes taking place during the final anneal tend to dominate. The effect is that after optimum annealing at the final diameter the superconducting properties of the homogenized samples become identical with those of the unhomogenized material.

Thus if the superconducting wire is subjected to heat treatment at the final diameter the application of a homogenizing anneal to the original ingots will be a useful operation, since the homogenization eliminates dendritic liquation and should lead to more uniform properties in long pieces of processed wire, which is very important in the winding of large solenoids.

Addition of Carbon to the Alloys. The effect of adding carbon on the superconducting properties of the materials is of special interest. In order to understand the nature of superconductivity it is important to know how interstitial atoms affect the superconducting properties of various alloys. On the other hand, such elements as carbon and oxygen constitute inescapable impurities in alloys, and it is important to know to what extent they should be removed from the metal.

It was shown in [8] that the introduction of oxygen (up to 0.025%) and carbon (up to 0.02%) into a Nb–25%Zr alloy raised the uperconducting properties of this alloy. The effect of various additives on the properties of Nb–Ti alloys was studied in [9] and it was found that although carbon slightly increased the transformation temperature it reduced the critical current.

In this paper we present some results relating to the effect of up to 0.4% of carbon on the critical current of Nb–50%Ti.

The alloys were melted by the method indicated earlier. The carbon was introduced in the form of niobium carbide. Alloys containing 0.1, 0.15, and 0.4% C were prepared.

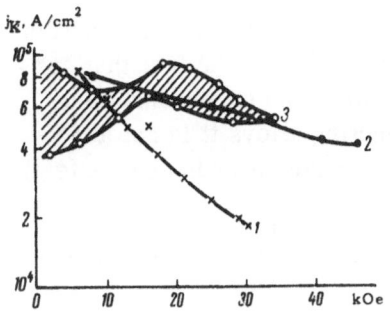

Fig. 6. Critical current density as a function of external magetic field for Nb–70%Ti samples. Final annealing temperature in °C; 1) 350; 2) 400; 3) 450.

The original ingots were hot-forged into bars and then cold-drawn into wire 0.25 mm in diameter. Samples with up to 0.15% C were quite ductile, but for larger amounts of carbon they became brittle. The critical current was measured in a transverse magnetic field of about 25 kOe.

Figure 4 shows the critical current in cold-worked wire samples as a function of carbon content. On introducing 0.15% C into the alloy the critical current density increases by a factor of about three. Further increasing the carbon content produces no further rise in critical current.

We studied the superconducting properties of Nb–50%Ti with 0.15% C after final annealing (at the final diameter). Figure 5 shows that the critical current density rises sharply as a result of annealing; after annealing at 350 and 450° the samples passed a current of 37 A without passing back into the normal state. There is a break in the graph, as the storage batteries of the apparatus were unable to supply a current over the range of the break.

The results may be explained on the basis of the Anderson theory [4]: The introduction of interstitial atoms into the solid solution causes additional distortions, which leads to a rise in the critical current.

Thus the presence of carbon increases the superconducting properties of the material both in the worked and in the annealed states. The introduction of carbon has a particularly considerable effect after final heat treatment. However, the carbon content should not exceed 0.1 to 0.15%, in order to avoid embrittlement of the wire.

Repeated Heat Treatment. We shall now present some data relating to the influence of final annealing, in combination with annealing at intermediate dimensions, on the superconducting properties of Nb–70%Ti.

We studied a semiindustrial batch of wire obtained from an ingot 5 kg in weight melted in an arc furnace with a consumable electrode in an argon atmosphere. As original materials we used niobium in block form and sheet titanium. The ingots were hot-forged into bars and then cold-drawn to produce wire 0.25 mm in diameter. In the course of drawing the wire was subjected to intermediate annealing (550° for 15 min) at diameters of 1.2, 0.8, and 0.55 mm. Annealing was carried out in a vacuum of 10^{-5} mm Hg.

At its final diameter of 0.25 mm the wire was annealed at 350, 400, and 450° for 1 h. Small measuring coils were wound from the wire (without applying any copper coating or insulation); the critical current was measured in these as a function of the external magnetic field.

We see from the results presented in Fig. 6 that the optimum heat treatment is annealing at 400°. The critical current density of a typical sample was $5 \cdot 10^4$ A/cm^2 in a magnetic field of about 50 kOe; a very important fact is that the curve relating j_K to H is shallow. The shaded region on curve 3 indicates the existence of a conditioning effect in these particular samples. We may well expect that the properties will be rather better for copper-coated and insulated samples.

Thus we see from the foregoing data that by choosing the appropriate heat treatment the critical current density in Nb–Ti alloys may be greatly increased.

A particularly effective method of raising j_K in Nb—Ti alloys is evidently the introduction of carbon (0.05 to 0.1%) in combination with final or intermediate annealing. The fact that carbon and oxygen [8] are not only not harmful but indeed beneficial toward the superconducting properties of the alloys means that in the production of superconducting alloys it is not absolutely essential to choose very pure original metals; the cheaper, standard brands are perfectly adequate. Our results confirm this conclusion.

In addition to this it is clear that the addition of carbon to the alloys will not only increase j_K but also stabilize the current in the solenoids.

LITERATURE CITED

1. T. G. Berlincourt and R. R. Hake, Phys. Rev. Lett., 9 (7):293 (1962).
2. J. K. Hulm and R. D. Blaugher, Phys. Rev., 123 (5):1569 (1964).
3. G. D. Kneip, J. O. Betterton, D. S. Easton, and J. O. Scarbrough, High Magnetic Fields, New York (1961), p. 603.
4. P. W. Anderson, Phys. Rev. Lett., 9:309 (1962).
5. C. J. Gorter, Phys. Lett., 1:69 (1962).
6. R. A. Kamper, Phys. Lett., 5:9 (1963).
7. W. Klose, Phys. Lett., 8:12 (1964).
8. J. O. Betterton, G. D. Kneip, D. S. Easton, and J. O. Scarbrough, Superconductors, New York (1962), p. 61.
9. V. V. Baron, M. I. Bychkova, and E. M. Savitskii, in: "Metallography and Metallophysics of Superconductors" [in Russian], Izd. "Nauka" (1965), p. 53.

EFFECT OF HEAT TREATMENT ON THE
SUPERCONDUCTING PROPERTIES OF V–Ti ALLOYS

Yu. V. Efimov, V. V. Baron, and E. M. Savitskii

The effect of annealing time and temperature on the critical current and superconducting-transformation temperature of V-Ti alloys is studied. Heat treatment raises the critical current of cold-worked alloys. The maximum critical current density at 26.6 kOe ($6.1 \cdot 10^4$ A/cm^2) is achieved in a V alloy containing about 74 wt.% Ti after annealing at 400° for 5 h. The critical temperature of the alloys changes much less. The effect of heat treatment on the superconducting characteristics is associated with the decomposition of the β-solid solution in the cold-worked alloys (degree of deformation 99%) during annealing under the influence of impurities.

In recent years the development of superconducting materials, magnets, and various devices based on these has occupied the attention of dozens of large companies and scientific organizations in the USA, England, and other countries [1, 2]. Superconducting wire is produced by several companies in the USA. Solenoids giving 70 to 100 kOe based on Nb–Zr and Nb–Ti alloys and Nb–Sn compounds have been devised [3]. Vanadium-titanium alloys also have a fairly high superconducting-transformation temperature (from 6 to 7.6°K for 30 to 70 wt.% Ti) [4, 5]. As regards the critical magnetic field V–Ti alloys are comparable with Nb–Zr [4-7]. The maximum critical magnetic field (about 110 kOe) occurs for an approximately equiatomic V–Ti alloy [6]. The critical current of V–Ti alloys is much lower than, for example, that of the Nb–Zr system, which has high values of $I_K(H)$; the maximum critical current density ($1.4 \cdot 10^4$ A/cm^2) at 26.6 kOe is reached for an alloy containing 50 wt.% Ti [8]. In our previous investigation [8] we showed that the critical current density of V–Ti alloys rose slightly after heat treatment.

The aim of the present investigation was to study the effect of heat treatment on the critical current and superconducting-transformation temperature of V–Ti alloys.

The original materials from preparing the alloys were iodide-type titanium (99.9% Ti) and carbothermal vanadium refined with cerium (99.766% V). The alloys were melted in an arc furnace with a nonconsumable tungsten electrode and a water-cooled copper base in an atmosphere of purified, titanium-gettered helium at a pressure of 0.7 atm. The homogeneity of the bars so produced (with respect to composition) was ensured by repeated remelting. The composition of all the alloys was checked by chemical analysis and in the majority of cases agreed fairly accurately with the calculated value.

The cast alloys were cold-forged and annealed at 800° for 1 h in a vacuum of 10^{-5} mm Hg. After annealing, the alloys containing up to 80 wt.% Ti were cold-rolled and drawn so as to obtain wire 0.2 mm in diameter (total degree of cold deformation 99%). The alloys with higher titanium contents were not cold-worked without intermediate annealing. The cold-worked samples, wrapped in tantalum foil with zirconium shavings, were annealed at 200 to 900° for 1 to 25 h and at 1100° for 1 h with water cooling. Heat treatment of the alloys was carried out in

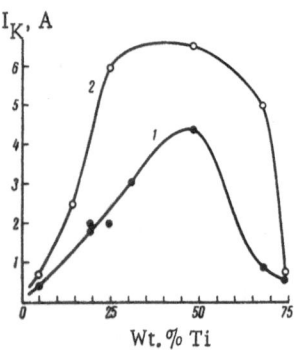

Fig. 1. Critical current of cold-worked (1) and 900°-annealed (2) V-Ti alloys.

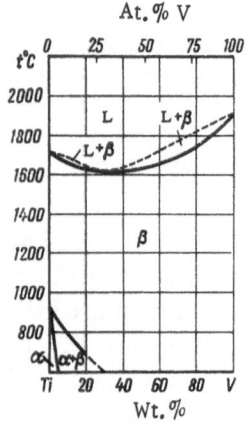

Fig. 2. Phase diagram of the V–Ti system.

double evacuated quartz ampoules. The annealed samples were selectively subjected to x-ray phase analysis.* X-ray pictures were taken directly from the wires in an RKD camera, using unflitered chromium radiation. In order to determine the lattice parameter the x-ray diffraction pictures were taken with the films inserted asymmetrically.

The critical current of the alloys was measured in liquid helium at 4.2°K, using short samples of diameter 0.2 mm and 100 to 120 mm long in a transverse magnetic field of up to 26.6 kOe. The critical current was determined three or four times for each value of the magnetic field until resistance appeared in the sample. The superconducting-transformation temperature was measured (for temperatures above 4.2°K) in a special apparatus constructed by N. D. Kozlova in the Laboratory of Rare and Refractory Metals of the Institute of Metallurgy. The samples constituted bars 10 mm long and 2 to 3 mm in diameter or bundles of 10 to 20 wires 0.2 mm in diameter of the same length.

For the maximum external magnetic field (26.6 kOe) the critical current of V-Ti alloys annealed at the final diameter (900° for 1 h) is higher than the critical current of cold-worked alloys (Fig. 1). A particularly considerable increase in critical current occurs for alloys containing 20 to 70% Ti. After annealing the cold-worked alloys at 750° for 1 h the rise in critical current is far less. Increasing the annealing period of alloys containing 15 to 70 wt.% Ti, slightly raises the critical current; however, the values achieved are smaller than after higher-temperature annealing (900°).

Brief annealing at 600° (1 to 5 h) produces hardly any change in the critical current of the cold-worked alloys. Increasing the annealing period at 600° to 25 h even leads to a fall in the critical current density of the alloys in the middle of the V-Ti system.

Our cast, water-quenched (from 1100°), and cold-worked alloys were single-phase and identical with vanadium in crystal structure (α-Fe type). After high-temperature annealing two phases appeared in the alloys containing 20 to 70% Ti. Both phases had a bcc lattice of the α-Fe type (a = 3.03 to 3.06 Å and 3.11 to 3.19 Å respectively). The proportion of the phase with the lower lattice parameter diminished on increasing the amount of titanium in the alloys (the proportions of the phases were determined qualitatively by reference to the changing intensities of the lines on the x-ray pictures). In alloys containing 5.15 and 74 wt.% Ti only one cubic phase was found after annealing. The slight fall in the lattice constant of the alloy containing 15% Ti also indicates the decomposition of the β solid solution in this alloy. The lattice constants of the β solid solution in alloys with 5 and 74% Ti remained the same after annealing (a = 3.03 and 3.21 Å respectively).

After annealing at 750 and 600° the x-ray diffraction pictures of the alloys in the middle part of the V–Ti system showed lines of two cubic solid solutions. However, the lattice parameters could not be determined owing to the severe blurring of the lines. Evidently the period

*The x-ray analysis of the samples was carried out specially by colleagues in the Inorganic Chemistry Faculty of the I. Franko L'vov State University, for which the authors are very grateful.

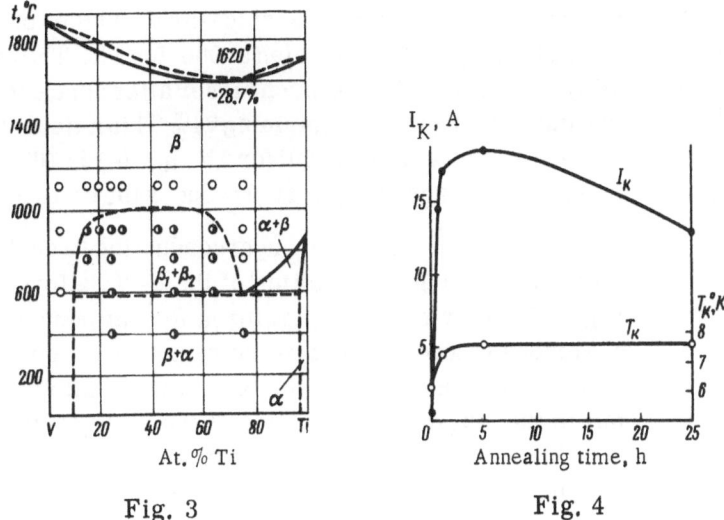

Fig. 3   Fig. 4

Fig. 3. Boundaries of the decomposition of the solid solution in the binary V-Ti system under the influence of impurities.

Fig. 4. Superconducting-transformation temperature and
critical current of a cold-worked vanadium alloy containing 74 wt.% Ti as a function of annealing time at 400°.

of annealing was insufficient to separate the phases completely. In alloys annealed at 400° a
hexagonal phase (a solid solution based on α-Ti) appeared as well as the cubic phase.

According to the phase diagram [9–11], vanadium forms a continuous series of solid solutions (Fig. 2) with the high-temperature β form of titanium. The low-temperature α form of
titanium existing below 882° dissolves up to 3.25 wt.% of vanadium. The two-phase region
($\alpha + \beta$) extends over a fairly narrow range of concentrations (from 3.25 to 21% V at 650°). The
observed decomposition of the β solid solution after the annealing of severely-worked alloys
containing 15 to 70 wt.% Ti is evidently associated with the presence of impurities in the alloys.
The approximate boundaries of the region of decomposition in the V—Ti system are shown in
Fig. 3. In alloys quenched from 1100° the β phase was always established in this range (15 to
70 wt.% Ti). This is completely legitimate, since the electron concentration of these alloys is
more than 4.2 electrons/atom [12]. On annealing quenched alloys the β solid solution decomposes under the influence of impurities into $\beta_1 + \beta_2$ or $\beta_1 + \alpha$, depending on the annealing temperature. A high degree of cold work (99% deformation) promoted the decomposition process.
Similar phenomena due to the influence of impurities were observed earlier [13] in Nb—Zr alloys.

The rise in the critical current of alloys containing 20 to 70% Ti is associated with the
decomposition of the β solid solution. Even annealing at 600° probably leads to the formation
of β_1 and β_2 solid solutions, the proportions of the components of these being close to the limiting values (10 and 75% Ti respectively). This reduces the distortion of the crystal lattices of
the phases and hence reduces the critical current. In addition to this, annealing at 600° evidently causes a reduction of the stresses in the cold-worked alloys.

The annealing of cold-worked alloys containing 25 and 75% Ti at 200° for 25 h has no effect
on the critical current. Brief annealing at 400° (0.5 to 1 h) causes a rise in the critical current
of the alloys (Fig. 4). This rise is evidently associated with the initial stage of decomposition
and the appearance of metastable intermediate phases, since an increase in the annealing period
leads to a fall in the critical current of the alloys. The maximum critical current in alloys with
25 and 48% Ti annealed at 400° occurs after holding for 0.5 h. The values of critical current

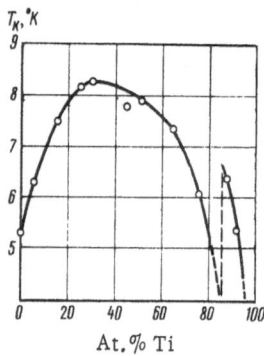

Fig. 5. Critical super-
conducting-transform-
ation temperature of
V–Ti alloys as a func-
tion of composition.

reached (6 to 7 A) are comparable with the values of critical current obtained in samples heat-treated at 900° for 1 h. The critical current of
alloys rich in titanium is much higher after annealing at 400°. The critical current of an alloy containing 74% Ti in a field of 26.6 kOe reaches
18.4 A after annealing at 400° for 5 h ($j_K = 6.12 \cdot 10^{-4}$ A/cm^2). Further
increasing the annealing period reduces the critical current (Fig. 4).

The variation in the superconducting-transformation temperature with varying composition of the V–Ti alloys is illustrated in
Fig. 5. The critical temperatures for cast and cold-worked alloys
of various compositions are the same for all concentrations. High-
temperature annealing of the cold-worked alloys leads to a certain
fall in the critical temperature of the alloy containing 25% Ti and a
rise in the critical temperature of the alloy containing 74% Ti. This
may be entirely explained by the decomposition of the β solid solution and the appearance of the V-richer β_1 solid solution in the
alloys. Annealing at 400° produces a sharper rise in the superconducting-transformation temperature of the alloy containing 74% Ti; this may also be explained
by the precipitation of a phase richer in vanadium. The considerable rise in the critical temperature of this alloy is associated with the large difference in vanadium content between the
original solid solution and the β solid solution precipitating as a result of annealing.

Conclusions

We have studied the effect of annealing time and temperature on the critical current and
superconducting-transformation temperature of V–Ti alloys. Heat treatment raises the critical
current density of cold-worked alloys. The maximum critical current density in a field of 26.6
kOe ($6.1 \cdot 10^4$ A/cm^2) is achieved in a vanadium alloy containing about 74 wt.% Ti after annealing
at 400° for 5 h. However, this value is no greater than the critical current density of the best
Nb–Zr alloys.

Heat treatment has a considerably smaller effect on the temperature at which V–Ti alloys
pass into the superconducting state.

The effect of heat treatment on the superconducting characteristics is associated with the
decomposition of the β solid solution of cold-worked V–Ti alloys in the course of annealing.
The decomposition of the cubic β solid solution into two cubic solid solutions on high-temperature annealing (or into a cubic solid solution with a higher vanadium content and a hexagonal
solid solution based on titanium as a result of low-temperature annealing) is due to the presence
of carbon and oxygen impurities in the alloys. A high degree of cold deformation (99%) applied
to the original alloys promotes decomposition.

LITERATURE CITED

1. E. M. Savitskii and V. V. Baron, Izv. Akad. Nauk SSSR, Metallurgiya i Gornoe Delo, No.
    5, p. 3 (1963).
2. E. Saur, Metall, 16 (5):380 (1962).
3. V. P. Kartsev, Superconductors in Physics and Technology [in Russian], Izd. "Znanie"
    (1965).
4. J. K. Hulm and R. D. Blaugher, Phys. Rev., 123 (5):1569 (1961).
5. D. K. Fox and W. Y. Reichennecker, Materials in Design Engineering, 57 (4):92 (1963).
6. T. B. Berlincourt and R. R. Hake, Phys. Rev. Lett., 9 (7):293 (1962).
7. U. Zwicker, Z. Metallkunde, 54 (8):477 (1963).
8. Yu. V. Efimov, V. V. Baron, and E. M. Savitskii, Collection, "Metallography and Metallo-
    physics of Superconductors" [in Russian], Izd. "Nauka" (1965), p. 59.

9.   H. K. Adenstedt, J. R. Pequinot, and J. M. Raymer, Trans. ASM, 44:990 (1952).

10.   W. Rostoker and A. Yamamoto, Trans. ASM, 46:1136 (1954).

11.   P. Pietrokowsky and P. Duwez, J. Metals, 4 (6):627 (1952).

12.   N. V. Ageev and L. A. Petrova, Dokl. Akad. Nauk SSSR, 138 (2):359 (1961).

13.   G. W. Berghout, Phys. Lett., 7 (1):292 (1962).

SUPERCONDUCTING PROPERTIES OF Nb—Hf ALLOYS
AND THE EFFECT OF MECHANICAL AND HEAT TREATMENT
ON THEIR STRUCTURE AND PROPERTIES

M. I. Bychkova, V. V. Baron, and E. M. Savitskii

The structure of Nb alloys containing 25, 50, and 75 at.% Hf is studied in the cold-worked and heat-treated states. The β solid solution decomposes into hexagonal and bcc phases at temperatures below 850°; above this temperature the solid solution decomposes into two bcc phases. The change in critical current and transformation temperature is studied for alloys after working and heat treatment. Heat treatment raises the superconducting characteristics; however, these are still not as high as those of Nb—Zr and Nb—Ti alloys.

Like Nb—Zr and Nb—Ti alloys, alloys of the Nb—Hf system are interesting from the point of view of their superconducting properties. Both metals are superconductors ($T_K^{Nb} = 9.4°K$; $T_K^{Hf} = 0.37°K$).

Niobium and hafnium (transition metals lying in neighboring groups) form a continuous series of solid solutions on interacting with each other at temperatures above the polymorphic transformation of hafnium. The high-temperature bcc solid solution decomposes below the polymorphic transformation point. Two forms of phase diagram have been published: According to [1] the decomposition of the solid solutions takes place in accordance with a monotectoid reaction over a wide range of concentrations; according to [2] the decomposition involves the reactions $β → β_1 → α$ in the same way as in the Nb—Ti system. In neither case was decomposition into two β-bcc solid solutions observed by x-ray diffraction. It was shown in [3, 4] that alloys of this system had quite high superconducting characteristics: a critical field of up to 80 kOe and a transformation temperature of about 10°K. However, these are lower than those of Nb—Zr alloys in both critical field and transformation temperature, and are considerable inferior to Nb—Ti alloys as regards the magnetic field.

In order to learn more about the properties of Nb—Hf alloys we studied those containing 25, 50, and 75 at.% Hf in the cold-worked state and also after annealing at various temperatures.

The alloys were prepared from niobium (electron-beam-melted) and hafnium (obtained by the iodide method) in an arc furnace with a nonconsumable electrode in an atmosphere of purified helium at a pressure of 0.5 atm. The ingots were cut into slabs 6 × 6 mm in cross section, rolled to 1.3 × 1.3 mm, and then drawn (through Pobedit tungsten-alloy dies with colloidal-graphite lubrication) to wires 0.25 mm in diameter.

The alloys were studied in the worked and annealed states. Annealing was carried out at 750, 850, 900, 950, 1000, 1050°. The alloys were annealed at the intermediate diameter of 1.3 mm (after which they were drawn to the final size) and at the final diameter of 0.25 mm. Heat treatment was carried out in quartz ampoules with subsequent water cooling. The annealing time was 1 h in each case.

74

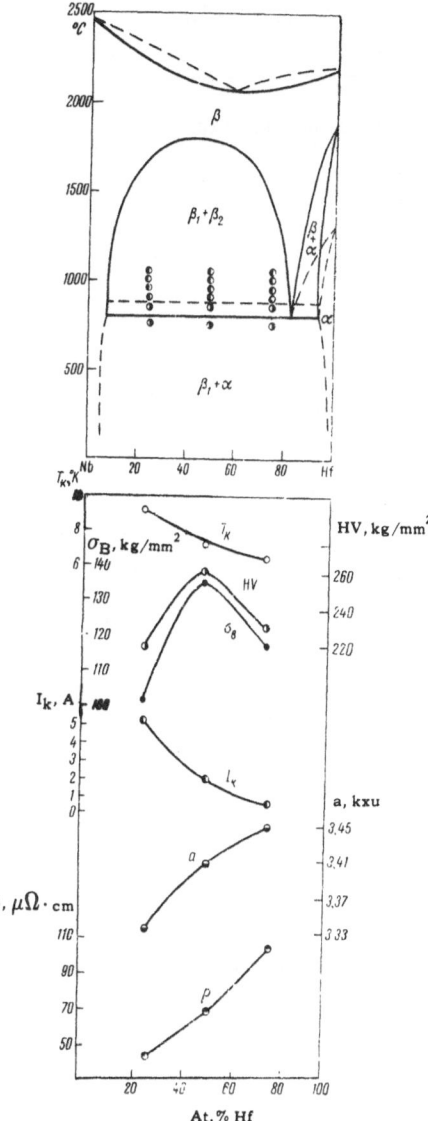

Fig. 1. Superconducting-transformation temperature, critical current, tensile strength, hardness, lattice constant, and electrical resistance as functions of the composition of Nb—Hf alloys.

A study of the structure of the cold-worked samples showed that the alloys were single-phase, with a lattice of the bcc type. The lattice constant increased on raising the proportion of hafnium in the alloys (Fig. 1).

After annealing at 750° the bcc lattice of the alloys with 25 and 50% Hf remained intact, but the lines on the x-ray picture were very diffuse. The lattice constant of the alloys with 25% Hf was unchanged but that of the alloy with 50% Hf fell from 3.41 to 3.37 kxu. In addition to the phase with the bcc lattice, the x-ray diffraction pictures of the alloy with 75% Hf showed the lines of a hexagonal phase; the lattice constant of the cubic phase was 3.42 kxu (Table 1).

Alloys of all three compositions annealed at 850° contain two phases: one with a bcc lattice and the other hexagonal. The parameter of the bcc lattice in the alloy with 25% Hf was unaltered after annealing at this temperature, remaining the same as in the cold-worked state. In alloys with 50 and 75% Hf the lattice constant of the bcc phase increased slightly after annealing.

After annealing at 950 and 1050° the original solid solution decomposes into two solid solutions with bcc lattices. In alloys with 50 and 75% Hf one of the cubic solid solutions formed has a lattice constant greater than in the original phase while the second has a smaller value. In the alloy with 75% Hf the lattice constant of one of the solid solutions formed approximately equals that of the original, while that of the second phase is much smaller.

We may conclude from these results that the phase diagram of the Nb—Hf system has a monotectoid decomposition, in agreement with the conclusions of [1]. The monotectoid line lies rather higher than in [1]: between 850 and 950°. The decomposition process taking place in the alloys has an effect on all their characteristics.

The critical current of the alloys studied in the cold-worked state is comparatively low. In the alloy with 25% Hf the critical current density equals $8.56 \cdot 10^3$ A/cm^2; with increasing Hf content this falls, and in the alloy with 75% Hf the critical current equals $8.14 \cdot 10^2$ A/cm^2.

Apart from the cold-worked state, the critical current was also measured in alloys subjected to intermediate annealing. The additional deformation of the annealed samples may slightly reduce the stresses introduced into the alloy on annealing. The critical current of the alloys subjected to intermediate annealing is higher than in alloys of the same compositions in the cold-worked state, (Table 2).

Thus the critical current in fields up to 26 kOe increases as a result of the decomposition of the solid solution into two bcc phases (annealing at 750° for 1 h) or into one bcc and one hexagonal phase. The highest critical current density was obtained in alloy with 25% Hf (intermediate annealing at 750°), namely, $1.6 \cdot 10^4$ A/cm^2, as compared with $8.56 \cdot 10^3$ A/cm^2 for the cold-

TABLE 1. Phase Composition of Nb—Hf Alloys According to X-Ray
Structural Analysis

Hafnium content, at.%	Quench temp. (after 99.8% cold work and annealing), °C	Phase composition and lattice constants, kxu
25	Cold-worked	bcc , $a = 3.34$
50	"	bcc , $a = 3.41$
75	"	bcc , $a = 3.50$
25	750	bcc , $a = 3.339$ *
50	750	bcc , $a = 3.37$ *
75	750	1) bcc , $a = 3.43$; 2) phase with hexagonal lattice
25	850	bcc , $a = 3.34$
50	850	1) bcc , $a = 3.42$; 2) phase with hexagonal lattice
75	850	1) bcc , $a = 3.42$; 2) phase with hexagonal lattice
25	950	1) bcc, $a = 3.34$; 2) bcc, $a = 3.29$
75	950	1) bcc, $a = 3.44$; 2) bcc, $a = 3.32$
25	1050	1) bcc, $a = 3.39$; 2) bcc, $a = 3.31$
50	1050	1) bcc, $a = 3.399$; 2) bcc, $a = 3.29$
75	1050	1) bcc, $a = 3.44$; 2) bcc, $a = 3.292—3.300$

* Severely blurred line on the Debye photograph

TABLE 2. Superconducting-Transformation Temperature and
Critical Current Density of Nb—Hf Alloys
as Functions of Treatment

HF content, at.%	Quench temp. (after 99.8% cold work and annealing), °C	Transformation temp. °K	Critical current density, A/cm²	HF content, at.%	Quench temp. (after 99.8% cold work and annealing), °C.	Transformation temp., °K	Critical current density, A/cm²
25	Cold-worked	9.6	$8.56 \cdot 10^3$	50	850	7.5	—
50	"	7.2	$1.4 \cdot 10^3$	50	900	7.5	—
75	".	6.1	$8.14 \cdot 10^2$	50	1000	7.5	—
25	750	9.7	—	50	1050	8.6—7.5	—
25	850	9.2	—	75	750	7.4	
25	900	9.1—7.5	—	75	850	7.5	—
25	950	9.3	—	75	900	7.1—6.3	—
25	1000	9.3—8.2	—	25	750*	—	$1.7 \cdot 10^4$
25	1050	—	—	50	750*	—	$6.12 \cdot 10^3$
50	750	7.4	—	75	950*	—	$5.3 \cdot 10^3$

* After preliminary 95% and subsequent (after annealing) 96.2% deformation.

worked alloys (i.e., it was doubled). The intermediate annealing increased the critical current density of the alloy with 50% Hf by a factor of 4.5, and that of the alloy with 75% Hf by a factor of 6 as compared with the cold-worked state.

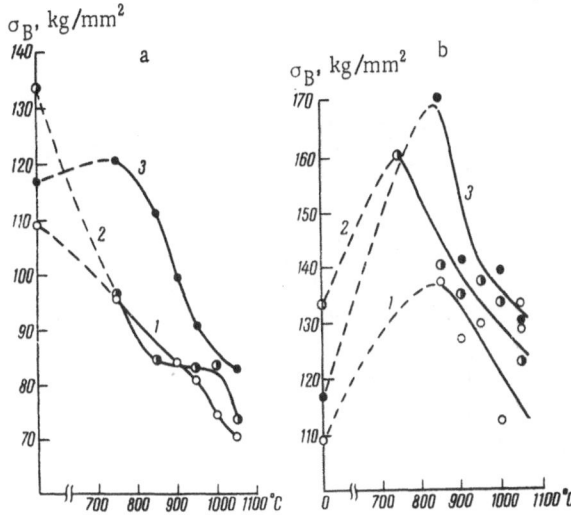

Fig. 2. Variation in the tensile strength of
Nb–Hf alloys with final (a) and intermediate
(b) annealing temperature. Hf content, at.%:
1) 25; 2) 50; 3) 75.

In the cold-worked state the highest super-conducting-transformation temperature occurred in the alloy with 25% Hf (9.4°K); on increasing the hafnium content the transformation temperature fell (Fig. 1). After annealing at 750° the transformation temperature of the alloy with 25% Hf rose slightly, then on raising the annealing temperature it fell to between 9.2 and 9.3°K; in addition to this, for the alloy with 25% Hf the transformation into the superconducting state occurred at a lower temperature (7.5°K and 8.2°K after annealing at 900 and 1000°C, respectively).

For an alloy with 50% Hf, after annealing at 750 to 1050° the temperature rose from 7.2 to between 7.4 and 7.5°K. For an annealing temperature of 1050° the temperature/induction curves showed two bends: one at 7.6 and the other at 8.6°K.

For an alloy with 75% Hf the transformation temperature increases after annealing; this occurs to the greatest extent after annealing at 750 and 850° (to 7.4 and 7.5°K respectively). In the alloy annealed at 900° there are two temperatures corresponding to the transformation into the superconducting state: 7.1 and 6.3°K.

The tensile strength and hardness of the alloys in the cold-worked state have a maximum at 50% Hf. This behavior is typical of properties in a region of unlimited solid solutions. The tensile strength of alloys subjected to both intermediate and final annealing differs considerably from that characterizing the cold-worked state.

On final annealing (sample diameter 0.25 mm) the tensile strength of alloys with 25 and 50% Hf falls below that of the cold-worked material for all annealing temperatures (Fig. 2). In the alloy with 75% Hf there is a maximum for an annealing temperature of 750°; at the remaining annealing temperatures the tensile strength is lower than that of the cold-worked state (Fig. 2).

On intermediate annealing the tensile strength/annealing temperature curves have a maximum at 750 to 850° for all three compositions (Fig. 2). In alloys with 25 and 75 at.% Hf the maximum tensile strength occurs at 850°. For alloys with 25 and 75% Hf the curves have a similar appearance; although after passing the maximum at 850° the tensile strength falls with any further increase in annealing temperature, the tensile strength of both alloys remains higher than that of the cold-worked material. In the alloy with 50 at.% Hf the tensile-strength maximum occurs at 750°. For an annealing temperature of 900 to 950° the tensile strength of the annealed alloy falls to the values characterizing the original cold-worked alloy, while for higher annealing temperatures the tensile strength of this alloy is even lower than that of the cold-worked material.

The specific electrical resistance (Fig. 3) of alloys with 25 and 50% Hf is little affected by the processes taking place in the alloys on either intermediate or final annealing.

The specific electrical resistance of the alloy with 75 at.% Hf changes considerably after annealing. A large change occurs on final annealing: at 750° a fall in specific electrical resistance as compared with the cold-worked alloy, and at higher annealing temperatures a rise. The curves are similar, although less sharply expressed, for intermediate annealing.

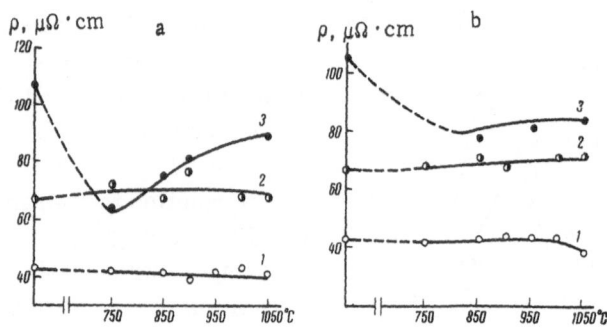

Fig. 3. Variation in the electrical resistance of Nb–Hf alloys with final (a) and intermediate (b) annealing temperature. Notation of curves as in Fig. 2.

By studying the properties of annealed Nb–Hf alloys, the phase-analysis data regarding the monotectoid decomposition taking place in the system are thus confirmed. This is particularly true of the transformation temperature. The fact that two transformation temperatures are obtained after annealing alloys with 25 and 75% Hf at 900°, both being well above 4.2°K, indicates the decomposition of this alloy into two cubic phases, each of which has its own transformation temperature. The proposition as to the precipitation of a hexagonal phase is discarded, since in the hexagonal phase the temperature is under 4.2°K and cannot be measured by our method.

Thus the determination of two transformation temperatures in alloys treated at 900 and 1050° (see Table 2) confirms the existence of the monotectoid decomposition, and also indicates that the monotectoid line lies below 900°. In Fig. 1 the broken line of the monotectoid is drawn between 850 and 900°, since at 850° x-ray structural analysis indicates the appearance of a new hexagonal phase (see Table 1), while at 900° the transformation temperature shows two cubic phases.

Conclusions

We have studied the superconducting properties of Nb–Hf alloys (25 to 75 at.% Hf) in the cold-worked state.

In studying the structure of the alloys after annealing (final and intermediate) we have established that the decomposition processes take place by a monotectoid type of reaction. We have calculated the parameters of the bcc lattice of the two solid solutions formed on decomposition.

The decomposition processes raise the critical current of the alloys and the transformation temperature as well. A rise in critical current occurs either when the alloy decomposes into two bcc solid solutions or when it decomposes into one phase with a bcc lattice and another phase which is hexagonal. The mechanism underlying the rise in current on decomposition is probably similar to that found in the analogous Nb–Zr system and also in Nb–Ti and V–Ti alloys, i.e., the precipitation of a less-superconducting phase in a superconducting matrix. It is possible that this process of the development of chemical inhomogeneities is accompanied by the creation of stressed regions around the precipitating phase.

Analysis of the resultant properties shows that the critical current in this system may be increased by securing still higher degrees of decomposition. The superconducting-transformation temperature may serve as a criterion in physicochemical analysis, revealing the presence of phases with different transformation temperatures in an alloy. In the present system this method should be at least as sensitive as x-ray analysis.

LITERATURE CITED

1. E. M. Savitskii, M. A. Tylkina, and I. A. Tsyganova, Zh. Neorg. Khim., 9(7):1960 (1964).
2. A. Taylor and N. J. Doyle, J. Less-Com. Metals, 7:37 (1964).
3. T. G. Berlincourt and R. R. Hake, Phys. Rev., 131:1 (1963).
4. J. K. Hulm and R. D. Blaugher, Phys. Rev., 123(5):1569 (1961).

III. THREE-COMPONENT SUPERCONDUCTING ALLOYS

STRUCTURE AND SUPERCONDUCTING PROPERTIES OF AN Nb—33%Zr ALLOY CONTAINING COPPER

V. A. Frolov, V. V. Baron, and E. M. Savitskii

The interaction of copper with an Nb—33%Zr alloy in the cast state and after vacuum-annealing at 1000° followed by rapid cooling is studied. The solubility of copper in this alloy at 1000° is under 0.3%. Copper present in an Nb—33%Zr in quantities up to 3.6% has practically no effect on the superconducting transformation temperature, and has a positive effect on the critical current.

Recently great attention has been paid to Nb—Zr alloys with high superconducting properties. The maximum superconducting-transformation temperature for these is around 11.3°K and the critical current density 10^4 to 10^5 A/cm^2 in fields of 50 kOe. Wire prepared from these alloys may be used to create solenoids with strong magnetic fields (of the order of 50 to 100 kOe). In our view it is particularly interesting to study the effect of copper on the structure and superconducting properties of Nb—Zr alloys. For these alloys copper is used as an insulating coating on wire wound into solenoids, and it is therefore important to discover how copper will interact with Nb-Zr alloys and how this will reflect on the superconducting characteristics.

There have not been very many investigations into the effect of metals with high electrical conductivity on superconducting metals and alloys (a few on binary systems, none on ternary). There are also no published phase diagrams or properties of the Nb—Zr—Cu system except for binary sections (Figs.1, 2, and 3) [1, 2]. On the basis of the phase diagrams of the binary systems we may suppose that a region of copper solubility adjacent to the Nb—Zr system should occur in the ternary system above 850 to 900°C.

In order to study the interaction of copper with niobium and zirconium in the ternary Nb—Zr—Cu system we took an alloy of Nb—33%Zr possessing good technological and fairly high superconducting properties.

The alloys were melted in an electric-arc furnace with a tungsten electrode on a water-cooled copper base in a medium of purified helium. Earlier test experiments on the fusion of copper with niobium showed that in the course of melting there was a great deal of evaporation of the copper (of the order of 85 to 90%); hence when preparing the mixture a quantity of copper 5 to 15 times greater than the calculated requirement was used.

The resultant alloys of Nb—33%Zr with copper contained (according to chemical analysis) 3.6, 0.5, 0.4, and 0.3% of copper and were uniform in composition. The microstructure of the alloys was studied and the hardness, superconducting-transformation temperature, and critical current density were measured.

In order to reveal the microstructure we used an etchant consisting of 75% HNO$_3$ and 25% HF. The hardness was measured on a Vickers hardness tester with a load of 30 kg. The

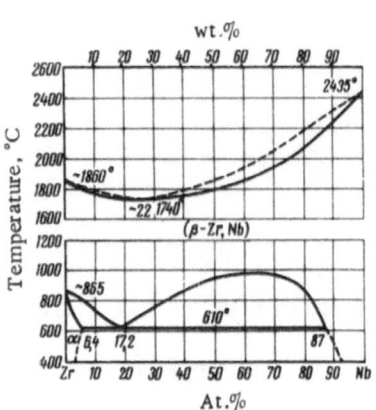

Fig. 1. Phase diagram of the
Nb–Zr system.

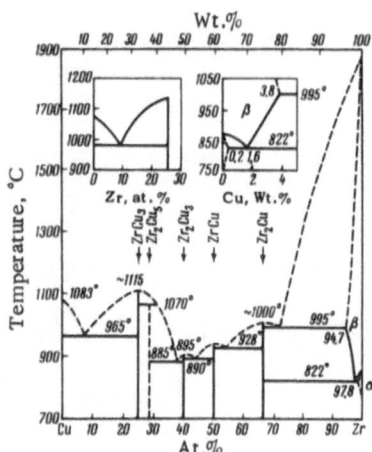

Fig. 2. Phase diagram of the
Zr–Cu system.

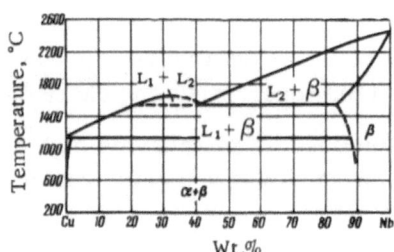

Fig. 3. Phase diagram of the
Nb–Cu system.

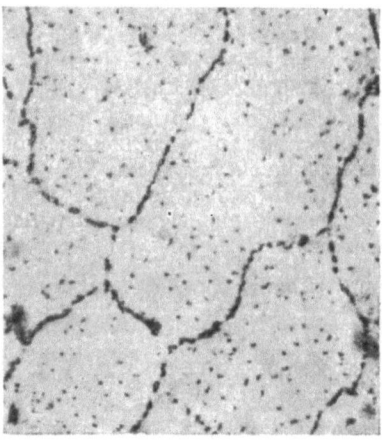

Fig. 4. Microstructure of a Nb–
33%Zr–0.3%Cu alloy (×500).

superconducting-transformation temperature and the critical current were determined in apparatus constructed by the A. A. Baikov Institute of Metallurgy.

In the cast state the alloys had a sharply-expressed dendritic structure. The hardness of the alloys containing copper differed little from that of the ordinary Nb–33%Zr alloy.

The alloys were annealed at 1000°C for 200 h in vacuum with subsequent rapid cooling. Metallographic analysis of the samples after annealing showed that particles of a second phase occurred in the structures of all the alloys (Fig. 4). The superconducting-transformation temperature of the annealed alloys containing copper differed little from that of the Nb–33%Zr. Thus, for example, in alloys with 3.6% Cu the transformation temperature was 10.8°K (in Nb–33%Zr it was 10.9°K). The critical current density of alloys containing 0.3 and 0.5% Cu measured on short samples of wire 0.2 mm in diameter (degree of deformation 99.8%) in a magnetic field of 26.6 kOe was

$5.4 \cdot 10^4$ A/cm^2 (in Nb–33%Zr with the same degree of deformation and the same magnetic field it equaled $3.8 \cdot 10^4$ A/cm^2).

Conclusions

1. The solubility of copper in Nb–33%Zr at temperatures of the order of 1000°C is less than 0.3%.

2. In the quantities studied copper hardly lowers the superconducting-transformation temperature of Nb–33%Zr at all; in fact, it has a beneficial effect on the critical current.

The structure of the Nb–Zr–Cu system on the side adjacent to the Nb–Zr system will be studied later and the effect of large amounts of copper on the superconducting characteristics of Nb–Zr alloys will be studied.

LITERATURE CITED

1. M. Hansen and K. Anderko, Structure of Binary Alloys [Russian translation], Vol. 2, Metallurgizdat (1961), pp. 699 and 1083.
2. G. V. Zakharova, I. A. Popov, et al., Niobium and Its Alloys [in Russian], Metallurgizdat (1961), p. 213.

SUPERCONDUCTING PROPERTIES
OF NIOBIUM-BASE ALLOYS

B. G. Lazarev, O. N. Ovcharenko,
A. A. Matsakova, and V. T. Volotskaya

Cast ternary Nb–Zr–Ti alloys are studied by x-ray diffraction after various forms of annealing and also in the quenched state. The ranges of existence of the various phases in isothermal cross sections of the ternary phase diagram are established. The decomposition of the solid solution in alloys situated within a wide area of the center of the concentration triangle takes place at temperatures below 600°, with the precipitation of the β phase (bcc lattice) containing a large amount of niobium and the hexagonal α phase containing little niobium. The superconducting characteristics are measured for Nb–Zr–Ti and Nb–Ti alloys in the cast and worked states and also after the annealing of cast and worked samples. The rise in superconducting-transformation temperatures in cast alloys after annealing is explained by the precipitation of a new phase rich in niobium. The annealing of cast alloys has no effect on the critical magnetic field. Severe deformation only slightly increases this characteristic.

It is well known that for solenoids with a superconducting winding one usually employs Nb-base alloys such as Nb–Zr and Nb–Ti. Ternary Nb–Zr–Ti alloys have also been developed and these have found wide application in the manufacture of laboratory solenoids with magnetic fields up to 90 kOe (at 1.8°K).

In order to discover the nature of the superconductivity of these alloys the superconducting properties (T_K, I_K, H_{K_1}, H_{K_2}) were measured and the same alloys were also subjected to x-ray structural examination.

In this paper we shall set out the results of some measurements of the critical temperaature T_K, the critical magnetic field H_{K_2}, and the value of $\left(\frac{dH_{K_2}}{dT}\right)_{T_{K_2}}$ for ternary Nb–Zr–Ti alloys over a wide range of concentrations. The measurements were made on cast (quenched) and worked (drawn) alloys.

The critical temperature T_K was measured by reference to the reestablishment of electrical resistance in the sample on passing from the superconducting state back to the normal state. The value of $\left(\frac{dH_{K_2}}{dT}\right)_{T_K}$ for these alloys was determined from the displacement of the superconducting-transformation curve in a magnetic field. The critical magnetic field H_{K_2}, was measured at 4.2°K in pulsed magnetic fields, also by reference to the restoration of resistance on passing from the superconducting to the normal state.

The results of the measurements of T_K and H_{K_2} for samples of cast ternary alloys are shown in Figs. 1 and 2; Figs. 1a and 2a give a three-dimensional picture of the surfaces corres-

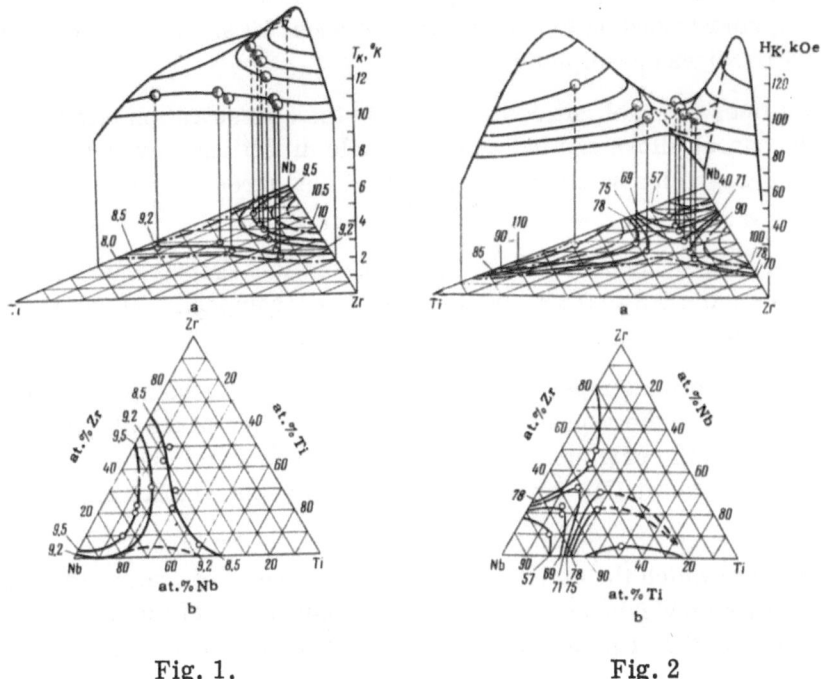

Fig. 1.   Fig. 2

Fig. 1. Three-dimensional picture of the surface of critical temperatures (a) and curves of equal values of T_K (b) for quenched
Nb–Zr–Ti ternary alloys.

Fig. 2. Three-dimensional picture of the surface of critical magnetic fields at 4.2°K (a) and curves of equal values of H_{K_2} at the
same temperature, for ternary alloys of the Nb–Zr–Ti system in
the quenched state.

ponding to the critical temperatures and critical magnetic fields, while Figs. 1b and 2b give
curves representing equal values of T_K and H_{K_2} on the concentration triangles. In order to
plot the graphs, published data relating to T_K [1] and H_{K_2} [2] for the binary alloys of the
Nb–Zr, Nb–Ti, and Zr–Ti systems were also used.

We see from Fig. 1a that the T_K surface for quenched Nb–Zr–Ti alloys is convex; it
rises from 9°K in the niobium corner to 11°K for Nb–25%Zr and then falls smoothly to 8.5°K
in the direction of the Ti–Zr side.

The H_{K_2} surface (Fig. 2a) is concave with a fairly wide plateau in the center of the concentration triangle, where H_{K_2} equals 70 to 80 kOe.

As indicated by x-ray structural analysis,* all the quenched ternary alloys were single-
phased with a bcc lattice (β phase).

The measurements of T_K and H_{K_2} may serve as initial data when studying the effect of
annealing on the superconducting properties of Nb–Zr–Ti alloys. It was found that the majority of these alloys (with a Zr content of over 10 at.%) underwent decomposition at 600°C or under,
with the precipitation of a Nb–rich bcc β' phase with a low-Nb hexagonal α phase; in all the de-

---

* The x-ray data and the measurements of critical magnetic field for these alloys are presented
  in another contribution to this volume: "Critical Magnetic Fields of High-Field Superconductors" (see p. 114).

composing alloys the β' phase had approximately the same lattice constant (independently of the original composition), corresponding to a Zr content of about 10 at.%, i.e., the limiting solid solution of Zr and Ti in Nb was precipitated.

Figure 3 shows lines of equal lattice constants on the concentration triangle for the original single-phase (β phase) alloys and indicates possible directions in which the composition of the bcc phase may change on annealing. These directions agree closely with the annealing-induced changes in T_K.

In the alloys involving decomposition of the solid solution, the superconducting-transformation temperature rises from 8.5 to 9.2°K, while in the nondecomposing alloys the critical temperature remains constant.

The higher transformation temperature in annealed alloys (as compared with those in the cast state) may be ascribed to the annealing-induced precipitation of the β' phase, which has a higher niobium content than the original β phase; the α phase, containing a lower proportion of niobium and having a T_K of about 1°K, never appeared in our experiments on the transformation temperature.

There were cases in which the critical temperature of an alloy undergoing decomposition remained unaltered. This may evidently occur if the temperatures of the superconducting transformation are the same for the original β phase and the precipitating β' phase of new composition.

The critical magnetic field H_{K_2} in the Nb–Zr–Ti alloys remained unchanged after annealing. After severe plastic deformation, for example, when preparing the wire, the H_{K_2} of these alloys rose slightly (by a few percent).

The value of $\left(\dfrac{dH_{K_2}}{dT}\right)_{T_K}$ equals –15 to 30 kOe deg for the ternary alloys over a wide range of concentrations in both the cast and worked states. The change in this quantity after annealing lies within the limits of measuring error.

Thus the ternary Nb–Zr–Ti alloys, like the binary Nb–Zr and Nb–Ti, tend to decompose. In order to achieve the optimum superconducting properties, these alloys should be subjected to such mechanical and heat treatment as lead to the precipitation of finely-dispersed phases in the alloy structure. The fine network of such second-phase particles is responsible for the high superconducting properties in the alloys, as shown by our earlier investigation [3] and in later work of other authors [4].

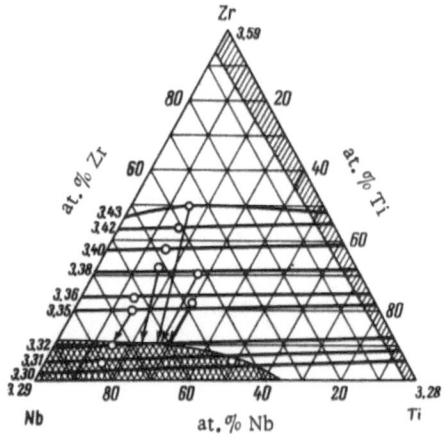

Fig. 3. Lattice parameters (in Å) of ternary Nb–Zr–Ti alloys in the quenched state (bcc β phase) [2]. The shaded parts of the concentration triangle provisionally denote the region of existence of the bcc β' phase (near the niobium corner) and the hexagonal α phase (near the Ti–Zr side).

LITERATURE CITED

1. J. K. Hulm and R. D. Blaugher, Phys. Rev., 123:1569 (1961).
2. C. K. Jones, J. K. Hulm, and B. S. Chandrasekhar, Rev. Mod. Phys., 36:74 (1964).
3. B. G. Lazarev, V. K. Khorenko, L. A. Kornienko, A. I. Krivko, A. A. Matsakova, and O. N. Ovcharenko, Zh. Éksp. Teor. Fiz., 45:2068 (1963).
4. E. P. Romanov, L. V. Smirnov, V. D. Sadovskii, and N. V. Volkenshtein, Fiz. Met. Metallov., 20:455 (1965).

THREE-COMPONENT ALLOYS BASED
ON THE Nb—Ti SYSTEM

M. I. Bychkova, V. V. Baron, and E. M. Savitskii

The effect of alloying with Zr, Hf, V, Ta, Mo, W, and Re on the structure and properties of a Nb–75%Ti alloy is studied. In the quenched, cold-worked state, these alloys constitute solid solutions with a bcc lattice. Zirconium and hafnium increase the lattice constant of the original alloy, Mo, W, and Re reduce it, while Ta has hardly any effect. All the additives in question reduce the critical current as compared with the original alloy in the cold-worked state. The transformation temperature is slightly higher in ternary alloys containing Zr, Ta, V, Mo, and W (up to certain percentages); rhenium reduces T_K.

In this contribution the effect of alloying with hafnium, molybdenum, and rhenium on the structure and superconducting properties (superconducting-transformation temperature and critical current) of a Ti–25%Nb alloy will be considered.

Molybdenum and hafnium form continuous series of solid solutions with niobium and the high-temperature form of titanium [1, 2]. It is already well known that rhenium also forms extensive series of solid solutions with niobium and high-temperature titanium [3, 4].

On the basis of existing binary phase diagrams (Nb–Ti, Nb–Re, Ti–Re) it is reasonable to expect a wide range of solid solutions with a bcc lattice in the ternary Nb–Ti–Re system, lying close to the Nb–Ti side of the concentration triangle.

The superconducting characteristics and the laws governing variations in these have been little studied for ternary systems of transition elements. Recently there has been one paper relating to the structure and superconducting properties of the Nb–Zr–Ti system [5].

Two-component Nb–Zr and Nb–Ti alloys with high superconducting parameters are already widely used in making superconducting magnets. A study of three-component alloys should help in the creation of new superconductors, possibly with higher characteristics than those of binary alloys, and should also aid our understanding of the laws governing changes in the superconducting properties of ternary alloys. The compositions of the alloys to be studied were chosen in such a way that they should lie in the single-phase regions of solid solutions with a bcc lattice. In alloying the samples the third component was introduced in place of titanium. The composition of the alloys is given in Table 1.

The original materials were metals of 99.7 to 99.8% purity. The original electron-beam-melted niobium contained (wt.%) 0.02 C, 0.03 N, and 0.01 O.

The alloys were melted in an arc furnace in an atmosphere of purified helium. The weight of the ingots was 40 g. These were homogenized at 1500° in a TVV-4 vacuum furnace. Then annealing took place at 1100° for 200 h with subsequent water-quenching. After this heat treatment, the alloys were subjected to 99.9% cold deformation.

90

TABLE 1. Composition, Specific Electrical Resistance, and Critical Current Density for Ternary Ti–Nb–Hf, Ti–Nb–Mo, and Ti–Nb–Re Alloys

Alloy composition, %			Specific electrical resistance ρ, $\mu\Omega \cdot$ cm	Critical current density j_K, A/cm^2	Tensile strength σ_b, kg/mm^2	Notes
Ti	Nb	Alloying additive				
75 1	24.9	—	92.7	$2 \cdot 10^4$	67.0	Cold-worked (99.9% deformation)
65	25	10/Hf	98	$2.4 \cdot 10^3$	—	"
50	25	25/Hf	83.2	$1.1 \cdot 10^3$	125.0	"
74,16	25.4	0.44/Mo	86	$3.3 \cdot 10^3$	88.0	"
69,6	25.4	5.0/Mo	80.5	$2.6 \cdot 10^3$	91.5	"
63,6	27.6	8.8/Mo	74.5	$1.4 \cdot 10^3$	110.0	"
46,25	27.7	26.05/Mo	—	—	—	Not cold-worked
74,0	25.0	1.0/Re	92.5	$3.45 \cdot 10^3$	92.5	Cold-worked (99.9% deformation)
70	25.0	5.0/Re	110.2	$1.4 \cdot 10^3$	110	"
65	25	10/Re	—	—	—	Not cold-worked
50	25	25/Re	—	—	—	"

After heat treatment the Ti–25%Nb alloy contained 0.058% O and 0.0011% H.

In order to determine the phase composition of the alloys we studied their microstructures in the cast and annealed states. The microsections were etched in an HNO$_3$–HF mixture. The Vickers hardness was also measured with a load of 30 kg.

In the cold-worked (99.9%) samples we determined the phase composition and lattice constants by the x-ray method, using an RKD camera and copper radiation. The accuracy in measuring the lattice constant by this method was ± 0.005 kxu.

The superconducting-transformation temperature T_K was measured in samples 7 mm long and not more than 3 mm in diameter (bundle of 99.9%-deformed wires). The accuracy of the measurement was ± 0.2°K.

The critical current of the alloys (wire of diameter 0.25 mm) was measured in a transverse magnetic field of up to 26 kOe at 4.2°K. The critical current was determined three or

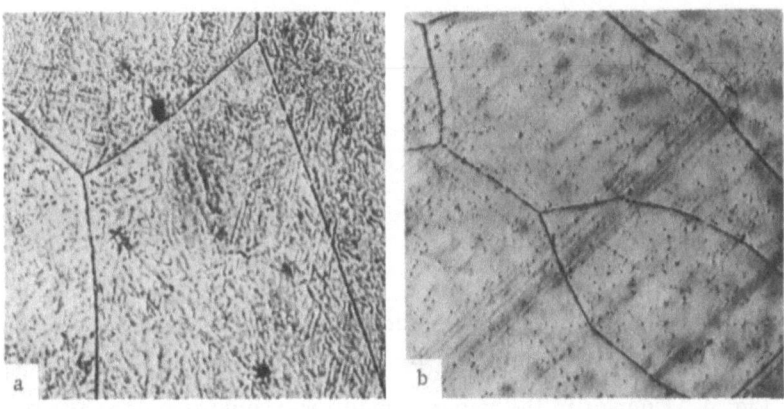

Fig. 1. Microstructures of cast, 1500°-homogenized, and 1100°-quenched Nb–75%Ti and Nb–65%Ti alloys (a and b respectively).

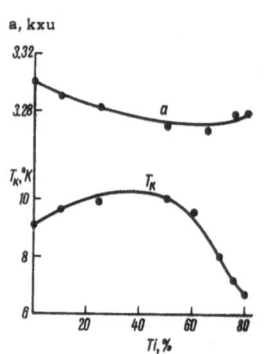

Fig. 2. Lattice constant and superconducting-transformation-temperature as functions of the composition of 99.9%-deformed Nb–Ti alloys.

four times for each value of the magnetic field. The accuracy of the measurements of critical current was 0.2 A.

The tensile strength of the wire 0.25 mm in diameter was determined on a miniature testing machine. The accuracy in determining the tensile strength was ± 3 kg/mm².

The specific electrical resistance was measured by a potentiometer method for the same wire. The accuracy of the measurement of electrical resistance was ± 2 $\mu\Omega \cdot$ cm.

X-ray diffraction showed that all the alloys were single-phased in the cold-worked state except for those containing rhenium. The x-ray pictures of the alloys containing rhenium showed a phase with a hexagonal lattice. We may suppose that the appearance of the second phase in these alloys was associated with the precipitation of a lower-temperature phase from the bcc solid solution as a result of solid-state transformations.

In the microstructural analysis of these alloys (homogenized, quenched state) no second-phase precipitates were found (Fig. 1), as in all the other alloys.

In the binary Nb–Ti system, on adding Ti to the Nb the lattice constant diminshed with a negative deviation from Vegard's law. On the composition/critical temperature curve (Fig. 2) of this system there is a maximum at 50% Ti (10°K). The lattice constant and transformation temperature of the Ti–25%Nb alloys were 3.28 kxu and 7.2°K respectively. The replacement of some of the Ti in this alloy by 25% Hf raised the lattice constant (from 3.28 to 3.34 kxu). Molybdenum and rhenium reduced the lattice constant of the original alloy with a negative deviation from the additive curve (Fig. 3). The transformation temperature fell in all cases. However, hafnium reduced the tranformation temperature very little; in the ternary alloy with 25% Hf the temperature was 6.9°K. The introduction of rhenium led to a sharp fall; in the ternary alloy with 25% Re the transformation temperature was under 4.2°K. Molybdenum in small quantities hardly affected the transformation temperature of the alloy; in an alloy with 5% Mo the transformation temperature was 7.4°K; however, further increasing the Mo content reduced the transformation point appreciably (Fig. 4).

In all cases alloying led to a rise in the tensile strength of the original Ti–25%Nb alloy. Moreover in alloys containing hafnium with the existing level of interstitial impurities a fair

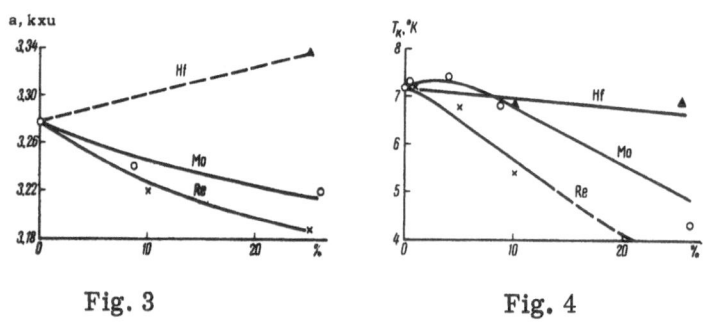

Fig. 3   Fig. 4

Fig. 3. Change in the lattice constants of a worked Nb–75%Ti alloy on adding Hf, Mo, and Re.

Fig. 4. Change in the superconducting-transformation temperature of a worked Nb–75%Ti alloy on adding Hf, Mo, and Re.

ductility was maintained over the whole range of concentrations. Alloys with molybdenum and rhenium (with the existing level of interstitial impurities) only retained a fairly high ductility on introducing up to about 9 and about 5% respectively. We see from the results presented in Table 1 that alloys with a tensile strength of up to 110 to 120 kg/mm^2 had a ductility enabling them to endure 99.9% cold deformation.

On introducing up to 10% Hf the specific electrical resistance of the alloys (at 20°C) increased; however, on further raising the Hf content the specific electrical resistance fell, and at 25% Hf it was lower than in the original alloy. Molybdenum and rhenium reduced the electrical resistance of the original alloy (Table 1).

All the alloying additives studied in the present investigation reduced the critical current density of the ternary alloys in comparison with the original Ti–25%Nb.

Conclusions

We have studied the structure, superconducting properties (transformation temperature, critical current), mechanical properties, and specific electrical resistance of ternary alloys obtained by alloying Ti–25%Nb with certain superconducting transition elements (hafnium, molybdenum, rhenium).

We have found that the superconducting properties characterizing the original Nb–Ti alloy are preserved in the ternary alloys constituting single-phase solid solutions with a bcc lattice.

All the ternary alloys studied have lower superconducting-transformation temperatures than that of titanium with 25% Nb. However, in ternary alloys including Hf this temperature remains quite high.

In Ti–Nb-base alloys containing Hf, Mo, and Re there is a correlation between the superconducting-transformation temperature and the lattice parameters; a fall in lattice parameter corresponds to a reduction in the superconducting-transformation temperature.

The ternary solid solutions studied evidently have a tendency toward an additive change of transformation temperature with changing alloy composition.

LITERATURE CITED

1.   H. J. Goldschmidt, J. Less-Com. Metals, 2:138 (1960).
2.   I. I. Kornilov and R. S. Polyakova, Zh. Neorg. Khim., 3(4):879 (1958).
3.   E. M. Savitskii, M. A. Tylkina, and K. P. Povorova, Rhenium Alloys [in Russian], Izd. "Nauka," Moscow (1965), p. 122.
4.   E. M. Savitskii, M. A. Tylkina, and Yu. A. Zot'ev, Zh. Neorg, Khim., 4(3):702 (1959).
5.   D. Tosio, J. Timihiko, and U. Tadasi, J. Japan Inst. Metals, 30(2):139 (1966).

EFFECT OF MICROINHOMOGENEITIES ON THE TRANSFORMATION TEMPERATURE OF SUPERCONDUCTING ALLOYS

I. A. Baranov, R. S. Shmulevich, V. A. Sytnikov, V. R. Karasik, and N. G. Vasil'ev

It is shown that in Nb–Zr–Ta and Nb–Ti–Ta alloys obtained by electron-beam melting the reduction of microscopic inhomogeneities (segregations) as revealed by x-ray microanalysis leads to a fall in the super-conducting-transformation temperature. The same effect is found in an Nb–52%Zr alloy after reducing the amount of microscopic segregation (liquation) by zone melting.

Earlier we studied alloys of the Nb–Zr–Ta system (based on 50 wt.% Nb) and found that the critical current increased in the presence of a high microstructural inhomogeneity [1].

In the present investigation we studied the effect of microinhomogeneities on the super-conducting-transformation temperature of 50%Nb–Zr–Ta, 50%Nb–Ti–Ta and Nb–52%Zr alloys.

At the same time it seemed particularly interesting to verify the effect of substituting some of the Nb in Nb–Zr and Nb–Ti alloys with its homolog Ta. From the point of view of its possible influence on T_K this substitution is the simplest: The concentration of the valence electrons per atom remains constant, and tantalum and niobium have very similar crystal-lattice parameters ($a_{Nb} = 3.294$ Å, $a_{Ta} = 3.296$ Å), so that the volume factor, which might otherwise affect T_K, may be neglected.

The method of preparing the samples of ternary alloys was described in [1] and that of Nb–52%Zr samples in [2]. The transformation temperature was measured by the induction method described in [3].

In view of the established fact that the superconducting-transformation temperature falls on reducing the microstructural inhomogeneity in ternary alloys, it seemed desirable to verify the effect of structural inhomogeneity on T_K for an Nb–Zr alloy subjected to zone refining in order to reduce the microinhomogeneity.

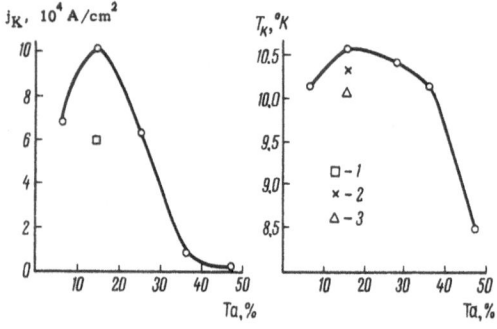

Fig. 1. Critical current density and super-conducting-transformation temperature of Nb–Zr–Ta (50 wt.% Nb) as functions of the Zr and Ta content. 1) Alloy with slight microinhomogeneity [1]; 2) after homogenization at 1200° for 8 h; 3) after further homogenization at 1500° for 4 h.

The composition of the Nb–52%Zr alloys was homogenized by the passage of a molten zone in an apparatus analogous to that described in [4], with a zone velocity of 20 to 25 mm/min. The sample was studied with an MAR-1 x-ray microanalyzer having a resolving power of 0.5 μ^2 and a limiting sensitivity of 0.05 to 0.2%. The homogenizing anneal was carried out in a TVV-4 furnace in a vacuum no higher than 10^{-5} mm Hg. The samples were previously wrapped in titanium foil.

Figure 1 shows the j_K and T_K for the group of Nb–Zr–Ta alloys studied. These alloys may be regarded as a system Nb–(0–50)%Zr in which some of the niobium has been replaced by tantalum. We see that the maximum value of T_K corresponds to an alloy with 50% Nb, 35% Zr, 15% Ta, for which the maximum current density is obtained.

This alloy was used to verify the effect of microinhomogeneity on T_K. Figure 2 shows the microstructure of the alloy samples in order of diminishing degree of liquation.

X-ray structural analysis showed that in both the cast and homogenized states the alloy had a single-phase solid-solution structure.

The results of our investigations show (Fig. 3) that there is a considerable fall in T_K with diminishing degree of liquation; the maximum T_K (10.5°K) corresponds to the cast state and the minimum (10.17°K) to the homogenized state.

Both in the cast and homogenized state the alloy shows a fairly sharp transition from the superconducting to the normal state. The width of the transition (ΔT_K = 0.1-0.3°K) is characteristic for alloys of such types existing in the single-phase state in the absence of external de-

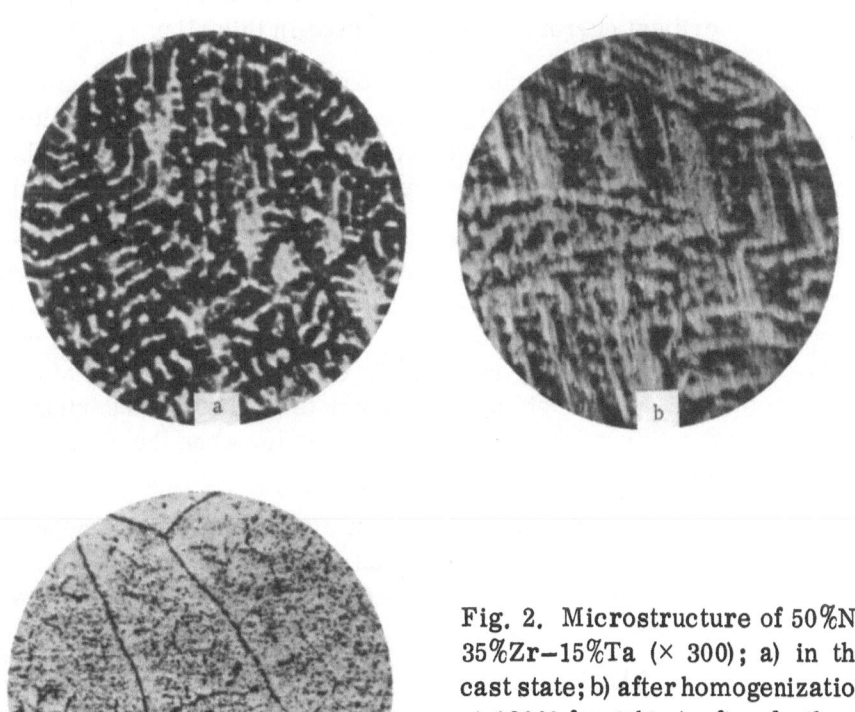

Fig. 2. Microstructure of 50%Nb–35%Zr–15%Ta (× 300); a) in the cast state; b) after homogenization at 1200° for 8 h; c) after further homogenization at 1500° for 4 h.

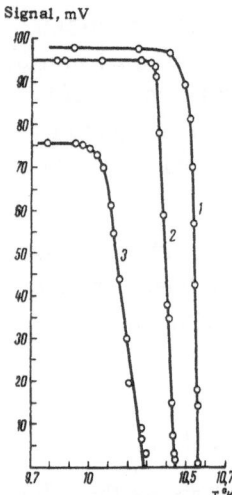

Fig. 3. Transition curves of a 50%Nb–35%Zr–15%Ta alloy for various degrees of homogenization. 1) Cast state; 2) after homogenization at 1200° for 8 h; 3) after further homogenization at 1500° for 4 h.

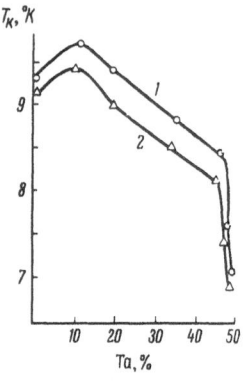

Fig. 4. Change in the superconducting transformation temperature of an Nb–Ti–Ta alloy containing 50 wt.% Nb as a function of the Ti and Ta content. 1) Cast state; 2) after homogenization at 1500° for 4 h.

fects and lattice distortions. As a result of homogenization the width of the transition increases slightly (from 0.15 to 0.28°K).

All the transition curves have practically the same slope, which indicates that there is no serious influence of the microinhomogeneities on the width of the transition.

The effect of the microinhomogeneity of an alloy on the values of T_K and ΔT_K was first observed in the case of Nb–Cr in [5]: In material with a well-developed dendritic liquation there was a rise in T_K, with a simultaneous sharp increase in the width of the transition ($\Delta T_K \approx 2°K$). The first phenomenon was explained as being due to the removal of chromium from the core, thus reducing T_K; as regards the rise in ΔT_K, this was regarded as being due to the continuous incorporation of parts of the alloy with different chromium contents into the superconducting transformation.

The absence of any rise in ΔT_K for an Nb–Zr–Ta alloy containing liquations indicates that the rise in T_K cannot in this case be attributed to fluctuations of composition. The following is the more likely mechanism: As the dendritic liquation process develops, not only the principal components, but also a number of impurities in the alloy undergo segregation. This leads to the formation of a framework of primary crystals free from impurities (and thus having a higher T_K) within the solidified material. The existence of a core or framework of this kind may serve as an explanation for the high critical current density observed in this alloy.

By comparing the T_K obtained in the ternary alloy (10.17°K) with the transformation temperature for the binary alloy Nb–35%Zr (10.6°K) we see that the replacement of some of the niobium with tantalum in Nb–Zr reduces the superconducting transformation temperature.

We also established a fall in transformation temperature for alloys of the 50%Nb–Ti–Ta system. We see from Fig. 4 that the replacement of some of the titanium by tantalum first raises T_K and subsequently reduces it. Over the whole range of alloys studied the transformation temperature falls after homogenization; however, the difference in T_K becomes smaller when the Ta content exceeds 46%.

In order to establish the effect of microinhomogeneities on the T_K of Nb–Zr alloys, samples of Nb–52%Zr were obtained with different degrees of microstructural inhomogeneity by zone melting. The effect of this process is shown in Fig. 5.

The degree of microstructural inhomogeneity with respect to a particular element may be characterized by a coefficient of microstructural inhomogeneity

$$K_{\text{Nb}} = \frac{I_{\max}^{\text{Nb}}}{I_{\min}^{\text{Nb}}} \text{ and } K_{\text{Zr}} = \frac{I_{\max}^{\text{Zr}}}{I_{\min}^{\text{Zr}}},$$

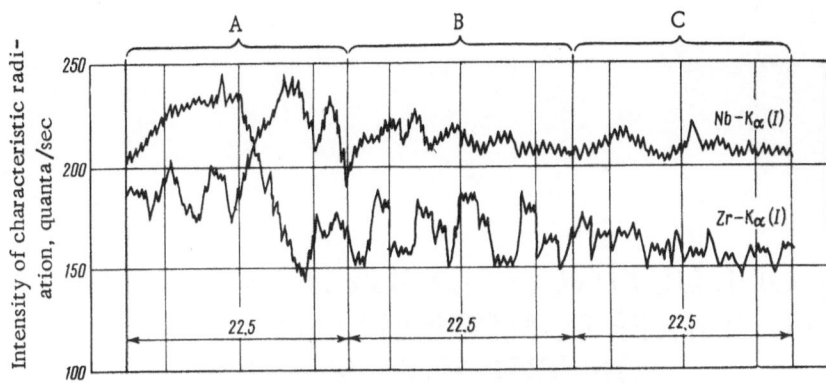

Fig. 5. Reduction in microinhomogeneities after zone-melting an Nb–52%Zr alloy in order to even out the components, expressed as a function of the number of zone passes. A) Original sample; B) after one zone pass; C) after two zone passes.

and for the alloy

$$K_{\text{Nb-Zr}} = \frac{K_{\text{Zr}}}{K_{\text{Nb}}}.$$

The general tendency for T_K to fall as a function of the degree of microinhomogeneity of the Nb–52%Zr alloy may be seen from the following results

	$K_{\text{Nb-Zr}}$	T_{K}, °K
Original alloy	1.35	9.63
After one zone pass . . .	1.26	9.46
After two zone passes . .	1.00	9.19

Thus our study of this group of alloys confirms the general tendency of the superconducting transformation temperature to fall as the microstructural inhomogeneities of the alloy diminish.

LITERATURE CITED

1. R. S. Shmulevich, I. A. Baranov, V. B. Novokreshchenova, V. R. Karasik, and G. B. Kurganov, Fiz. Met. Metalloved., 21(3):379 (1966).
2. I. A. Baranov, R. S. Shmulevich, V. R. Karasik, and G. B. Kurganov, in: "Metallography and Metallophysics of Superconductors" [in Russian], Izd. "Nauka," Moscow (1965), p. 72.
3. S. Sh. Akhmedov and T. N. Vylegzhanina, ibid., p. 120.
4. E. M. Savitskii et al., Zavod. Lab., No. 8, p. 957 (1961).
5. J. K. Hulm and R. D. Blaugher, Phys. Rev., 123(5):1569 (1961).

SUPERCONDUCTING AND MECHANICAL PROPERTIES OF THREE-COMPONENT ALLOYS BASED ON THE V–Ti SYSTEM

Yu. V. Efimov, V. V. Baron, and E. M. Savitskii

The effect of Zr, Hf, Nb, Ta, Mo, W. Re, Al, and Sn (up to 10 at.%) on the microstructure, hardness, ductility, strength, superconducting-transformation temperature, and critical current of cold-worked V alloys containing 25 to 85 at.% Ti obtained in an arc furnace with a helium atmosphere is considered. On increasing the amount of the alloying elements the critical current, transformation temperature, and ductility of the equiatomic V–Ti alloy are reduced and the strength and hardness increased. There is a particularly sharp fall in superconducting characteristics on introducing nontransition metals. A rise in superconducting properties occurs on alloying certain nonequiatomic binary alloys.

Alloys of the V–Ti system are comparable with Nb–Zr alloys as regards critical magnetic field [1]; however, they have much lower critical currents and superconducting transformation temperatures.

The aim of the present investigation was to study the effect of alloying on the superconducting transformation temperature and the critical current of V–Ti alloys.

The basis for the majority of ternary alloys studied was an equiatomic binary V–Ti alloy; the properties of this alloy were studied earlier [2]. As alloying elements we took transition metals of the second and third long periods of Groups IV to VII of the periodic system (Zr, Hf, Nb, Ta, Mo, W, and Re), and also (for comparison) two metals of the principal subgroups of the periodic system (Al, Sn) readily soluble in the original components [3]. This choice of alloying elements was made in order to discover the effect of the position of the alloying element in the periodic system on the superconducting characteristics of V–Ti alloys.

In all the alloys studied the amount of alloying elements equalled 10 at.%. All the alloying elements were introduced as replacements for titanium, in such a way that the amount of vanadium in the alloy was equal to the total content of titanium and the third component. In addition to this, alloys were prepared with a constant ratio of vanadium to titanium (3:1, 1:3, 1:1, and 3:17).

The original materials were metals 99.6 to 99.95 wt.% pure. The original electrolytic vanadium contained 0.005 C, 0.01 N, and 0.01 O (in wt.%). All the alloys were melted in an arc furnace on a water-cooled copper hearth in an atmosphere of purified, titanium-gettered helium. The most easily-melted and most refractory additives (Al, Sn, Ta, W, Re) were introduced as ligatures (Ti + 50 wt.% additive). In preparing the charges for the alloys with volatile additives (Al, Sn) allowance was made for the possibility of evaporation, this being determined in preliminary tests. The composition of the resultant alloys was verified by check weighing and selective chemical analysis. For the majority of alloys the deviations from the specified composition were no greater than ± 1%.

98

The microstructure and hardness of the alloys were studied in the cast state. The micro-sections were prepared by the ordinary methods of grinding and polishing and then etched in a mixture of HF and HNO_3 (19:1). The hardness was measured by the Vickers method with a load of 5 kg, using electropolished samples. The cast alloys were forged at 400 to 1200° and cold-drawn without intermediate annealing into wire 0.2 mm in diameter (the drawing lubri-cant was dried aquadag). The total degree of cold deformation was over 99%. On cold working, the V–Ti alloys containing 10 at.% of W, Hf, Re, Ta, or Sn, ruptured for degrees of deforma-tion exceeding 10 to 30%. In order to prepare wire from these alloys hot working was employed or in some cases cold working with intermediate annealing.

The superconducting transformation temperature was measured on samples 7 to 10 mm long and 2 to 2.5 mm in diameter cut from the cast alloys. The accuracy of measuring the critical temperature was ± 0.2°K. The critical current of the alloys was measured in liquid helium at 4.2°K, using short cold-worked samples 0.2 mm in diameter and 250 mm long in a transverse magnetic field of up to 26 kOe. The critical current was determined three or four times for each value of the magnetic field until resistance appeared in the sample. The accuracy of measuring the critical current was ± 0.2 A.

The original binary alloy and the ternary alloys containing 10 at.% of Ta, Nb, W, Mo, or Re were single-phased in the cast state. In these alloys x-ray diffraction revealed one solid solution with a bcc lattice. The lattice parameters of the alloys were not calculated owing to the diffuseness of the lines on the x-ray pictures, apparently due to stresses developing in the alloys on cooling after melting (partial quenching).

On microstructural analysis of the ternary alloys containing 10 at.% zirconium, hafnium, aluminum, or tin, finely-dispersed second-phase precipitates were found. The nature of these precipitates indicated that the alloys transformed into a single-phase state at high tempera-tures. The appearance of precipitates in these cast alloys was evidently due to the sharp re-duction in the solubility of the alloying elements in an equiatomic V–Ti solid solution with fall-ing temperature. The results of microstructural analysis agreed with the well-known ternary phase diagrams of the V–Ti–Mo (W, Nb, Ta, Zr, Al, Sn) systems [4-9]. The existence of a ternary solid solution containing 10 at.% Re agreed with the high solubility of Re in V [10] and Ti [3]. The solubility of Hf in a binary V–Ti alloy (as in V [11]) was under 10 at.%.

Tantalum, niobium, molybdenum, and zirconium change the hardness of the original alloy comparatively little. Elements further removed from the original alloy components in the periodic system cause a comparatively large increase in the hardness of the alloy. As a mea-sure of the extent to which they increase the hardness of the cast alloys (for 10 at.% of the additive) the alloying elements may be placed in the following ascending order: Nb, Ta, Mo, Zr, Hf, Al, W, Re, and Sn.

All the alloying elements raise the tensile strength of the binary V–Ti alloys. As a mea-sure of the extent to which they increase the tensile strength of an equiatomic V–Ti alloy the transition metals may be placed in the following order: Zr, Nb, Re, Mo, Hf, Ta, and W.

The critical temperature of V–Ti alloys with 25, 50, and 75 at.% Ti prepared from elec-trolytic vanadium is slightly higher than that of analogous alloys prepared from less pure materials; it equals 7.2, 7.8, and 6.2°K respectively. The critical temperature of V–85%Ti is under 4.2°K.

On taking the place of Ti in an equiatomic V–Ti alloy, all the alloying elements reduce the superconducting transformation temperature. The critical temperature falls particularly sharply on introducing nontransition elements (Al and Sn) into the alloys. Regarding the extent to which they reduce the critical temperature for 10 at.% of the additive, the third elements may be placed in the following order: Nb, Zr, Ta, Mo, Hf, W, and Re; the first of these reduces the transformation temperature by tenth parts of a degree and the last by whole degrees. In this series the critical temperature of the alloy is the lower, the further the alloying element is from the original components in the periodic system, i.e., the greater the difference in the structure of the electron shells of the alloying element and the original components of the binary alloy. No direct relationship was established between the critical temperature of the equiatomic V-Ti alloy and the change in electron concentration on alloying. The rise in the electron concentration of the alloy on replacing titanium by transition metals of Groups V to VII is accompanied by a fall in critical temperature, but so is the replacement of titanium by zirconium or hafnium, in which the electron concentration remains constant.

The difference in the atomic radii of the alloying elements and titanium (taken from [12]), although causing a change in the unit-cell volume of the solid solution, had no unambiguous effect on the critical temperature of the equiatomic V-Ti alloy. The addition of Zr, Hf, Ta, and Sn increased the size of the unit cell, while the addition of Mo, W, Re, and Al reduced it; however, the critical temperature of the equiatomic alloy fell in both cases.

A slight rise (less than $0.5°K$) in the critical temperature was observed on alloying non-equiatomic V-Ti alloys with transition metals.

Clearly the best conditions for a maximum critical temperature are realized in the binary equiatomic V–Ti alloy. Any change in the electron concentration of the original binary alloy on introducing transition metals reduces the critical temperature, even a very slight reduction in electron concentration leading to a sharp fall in the latter. On raising the electron concentration the fall in critical temperature is proportional to the change in electron concentration. On introducing various amounts of transition metals of Groups V to VII, the extent to which the critical temperature is reduced increases with the Group number of the alloying element, i.e., with increasing rise in electron concentration. However, even when the titanium is replaced by zirconium or hafnium (leaving the electron concentration unchanged), the critical temperature of the alloy still falls. This indicates that in this case other factors must affect the critical temperature. One notices a sharper change in the critical temperature of the alloy on replacing titanium with aluminum or tin, i.e., elements differing more sharply from the original components in the structure of their electron shells.

Measurements of the critical current of cold-worked ternary-alloy wires in magnetic fields up to 26 kOe showed that all the alloying additives studied reduced the critical current of the binary V–Ti alloy. Either without a magnetic field or after application of the latter, the critical current of the ternary alloys was considerably below the corresponding current for the binary V–Ti alloys. Alloying with nontransition metals caused a particularly sharp drop in critical current.

LITERATURE CITED

1. T. B. Berlincourt and R. R. Hake, Phys. Rev. Lett., 9(7):293 (1962).
2. Yu. V. Efimov, V. V. Baron, and E. M. Savitskii, in: "Metallography and Metallophysics of Superconductors" [in Russian], Izd. "Nauka," Moscow (1965), p. 59.
3. M. Hansen and K. Anderko, Constitution of Binary Alloys, McGraw-Hill, New York (1958).
4. I. I. Kornilov and V. S. Vlasov, Zh. Neorg. Khim., 5(7):2017 (1960).
5. A. Nowikow and H. J. Baer, Z. Metallkunde, 49(4):195 (1958).
6. S. Komjatby, J. Less-Com. Metals, 3(6):468 (1961).
7. E. Ence, F. A. Farrer, and H. Margolin, Binary and Ternary Diagrams, New York Univ. Coll. Engng., New York (1960), p. 82.
8. X. Koster and K. Hang, Z. Metallkunde, 48(6):327 (1957).
9. I. I. Kornilov and R. S. Polyakova, Izv. Nauk SSSR, Otd. Tekh. Nauk., No. 4, p. 76 (1961).

10. M. A. Tylkina, K. B. Povarova, and E. M. Savitskii, Zh. Neorg. Khim., 5(8):1907 (1960).
11. R. Kieffer and H. Braun, Vanadin-Niob-Tantal, Berlin (1963).
12. E. Teatum, K. Gschneider, and J. Waber, Los Alamos Scient. Lab. Report No. 2345 (1960).

IV. SUPERCONDUCTING PROPERTIES OF COMPOUNDS AND COATINGS PREPARED FROM THEM

THIRD CRITICAL FIELD OF SUPERCONDUCTORS
WITH A FILM ON THE SURFACE

V. V. Shmidt

The third critical field of a system comprising a thin superconducting film deposited on the flat surface of a massive superconductor is calculated. It is considered that the film differs from the substrate simply in respect of the mean free path of the electrons. Two limiting cases are considered: those of a thin and a thick film respectively. A qualitative explanation is given for the experimentally-observed dependence of the third critical field of a superconductor on the mechanical and heat treatment of the surface.

The existence of a third critical field in superconductors (H_{c3}) was predicted theoretically by Saint-James and Gennes [1] and later this effect was observed experimentally by a number of research workers. The essence of the matter lies in the fact that, if a superconductor of the second kind, having a flat surface, is placed in a very strong magnetic field parallel to the flat surface (for which it is known to be in the normal state) and the field is then reduced, at a certain field $H_{c3} = 1.695\,H_{c2}$ (H_{c2} being the second critical field) a thin superconducting layer will appear on the surface.

Recently there have been a number of papers in which this phenomenon has been studied experimentally. However, the experimentally-observed H_{c3}/H_{c2} ratio has not always been very close to the theoretical. Let us give some examples.

Well-annealed Nb–Ta alloys gave $H_{c3}/H_{c2} = 1.71$ [2]. This ratio was measured for an In–Pb alloy in [3]. It was found that for electropolished samples annealed at room temperature $H_{c3}/H_{c2} = 4$; after etching the ratio fell to 3, and after 60-h annealing at 110°C it became $1.69 \pm 10\%$. Annealed and later electropolished samples of $Sn_{0.62}Tl_{0.38}$ gave $H_{c3}/H_{c2} > 3$; this value fell sharply after holding for a week at room temperature, but rose again after further electropolishing [4].

These examples show convincingly that the surface quality has a great influence on the value of the third critical field. By "quality" we naturally mean the amount of impurity, the lattice distortions, the number of dislocations, etc. in the surface layer of the sample. A reduction in the mean free path of the electron is the common feature in all these factors.

Presentation of the Problem

The problem of the present investigation is to find H_{c3} in the case in which the plane surface of a massive superconductor is "contaminated" to a certain depth by atoms of another type, dislocations, and distortions which reduce the mean free path of the electron. In other words, we have the problem of finding H_{c_3} for a system comprising a massive superconductor with a superconducting film deposited on its surface, the material of this film having the same critical temperature and critical thermodynamic field as the massive substrate material but (according to Gor'kov [5]) a higher Ginzburg-Landau parameter $\varkappa$ than that of the underlying material [6].

Let the parameter $\varkappa$ of the massive substrate material be $\varkappa_2$. On the surface we have a film of thickness d with a parameter $\varkappa_1$ (in future the index 1 will relate to the film and 2 to the substrate). The plane boundary between the film and vacuum coincides with the plane $z = 0$. The film and substrate occupy the half space $z > 0$. All the electron characteristics of materials 1 and 2 are the same apart from the electron mean free paths, which respectively equal l_1 and l_2. A very high external magnetic field H_0 is applied parallel to the surface of the film, such that the whole system is in the normal state. Let us start reducing the field H_0. We have the problem of finding the field for which superconductivity starts developing in the surface layer. This field we shall call H_{c_3} as before.

Since we shall be confining attention to the neighborhood of T_c (critical temperature), we may solve the problem by the Ginzburg-Landau method [6]. The development of surface superconductivity at a field H_{c_3} is a phase transformation of the second kind, the parameter of the transformation being Ψ, the effective wave function. Near H_{c_3}, $|\Psi| \ll |\Psi_\infty|$, where $|\Psi_\infty|^2$ is the density of the superconducting electrons in the massive material at the specified temperature but without the magnetic field. In view of this inequality we may omit the term cubic in Ψ in the Ginzburg-Landau equation for Ψ and thus linearize the problem.

If we further select the vector potential A in such a way as to make Ψ real (using the property of the gradient invariance of the equations), the original system of equations will take the form

$$-\frac{\hbar^2}{2m}\frac{d^2\Psi_i}{dz^2} + \frac{2e^2}{mc^2} A^2\Psi_i + a_i\Psi_i = 0, \tag{1}$$

$$\frac{d^2A}{dz^2} - \frac{16\pi e^2}{mc^2}\Psi_i^2 A = 0, \quad i = 1, 2 \tag{2}$$

(as already indicated, the index 1 corresponds to the film and 2 to the substrate). Here a is the coefficient in the expansion of the free energy in powers of $|\Psi|^2$ ($F_s = F_{n_0} + a|\Psi|^2 + \frac{b}{2}|\Psi|^4$), e is the charge on the electron, m is the effective mass of the electron, and c is the the velocity of light.

The boundary conditions of the problem are

$$H|_{z=0} = H_0, \quad \frac{d\Psi}{dz}\Big|_{z=0} = 0, \quad \Psi|_{z=\infty} = 0. \tag{3}$$

In addition to this we still have to impose the conditions for the matching of the functions Ψ_1 and Ψ_2 at the interface between materials 1 and 2. For this purpose at $z = d$ we apply the boundary conditions for Ψ and $d\Psi/dz$ obtained by Zaitsev [6]:

$$\frac{1}{\sqrt{\chi_1}}\Psi_1 = \frac{1}{\sqrt{\chi_2}}\Psi_2, \quad \sqrt{\chi_1}\frac{d\Psi_1}{dz} = \sqrt{\chi_2}\frac{d\Psi_2}{dz}.$$

Here χ is a dimensionless function of the electron mean free path as derived by Gor'kov [5]. The function equals unity for $l = \infty$ and tends to zero as $l \to 0$. The Ginzburg-Landau constant $\varkappa$ for the alloy and $\varkappa_0$ for the pure metal (the alloy differs from the pure metal in respect to the electron mean free path only) are related by

$$\varkappa = \frac{\varkappa_0}{\chi}. \tag{4}$$

In the zero approximation $\Psi = 0$, and equation (2) is solved with due allowance for boundary conditions (3):

$$A = H_0 (z - z_0), \tag{5}$$

where z_0 is an integration constant.

Let us simplify equation (1) by transforming to dimensionless units. As in the case of Abrikosov [8], we take the unit of length as $(\hbar c/2eH_0)^{1/2}$ and transform to the dimensionless length ξ:

$$\xi = \sqrt{\frac{2eH_0}{\hbar c}}\, z.$$

We note that this length depends on the magnetic field only, not on the sample material.

Let us move the origin of coordinates to the point $z = z_0$. Then equation (1) takes the simpler form

$$\frac{d^2\Psi_i}{d\xi^2} + (\beta_i - \xi^2)\,\Psi_i = 0, \tag{6}$$

where

$$\beta_i = \frac{m\,|\,a\,|\,c}{e\hbar H_0} = \frac{\sqrt{2}\varkappa_i H_{\text{КМ}}}{H_0} = \frac{H_{c2}(i)}{H_0}. \tag{7}$$

Here $H_{\text{КМ}}$ is the critical thermodynamic field and H_{c_2} is the second critical field of the i-th material.

The boundary conditions and the matching conditions are

$$\frac{d\Psi_1}{d\xi}\bigg|_{\xi=-\eta} = 0, \quad \Psi_2|_{\xi\to\infty} = 0, \quad \sqrt{k}\,\Psi_1|_{\xi=s-\eta} = \Psi|_{\xi=s-\eta}\,,$$

$$\frac{1}{\sqrt{k}}\frac{d\Psi_1}{d\xi}\bigg|_{\xi=s-\eta} = \frac{d\Psi_2}{d\xi}\bigg|_{\xi=s-\eta}\,. \tag{8}$$

Here s and η are the distance z_0 and d expressed in the dimensionless units; we have also introduced the notation k = $\varkappa_1/\varkappa_2$.

The general solution of equation (6) contains four arbitrary constants of integration (two for Ψ_1 and two for Ψ_2). The four conditions (8) enable us to eliminate these constants and establish a functional relationship between the parameters β_i and η. The critical field for the development of surface superconductivity H_{c_3} will correspond (see [8]) to the maximum value of the external field H_0 for which there is a nontrivial solution of equation (6). As in [8] this field is determined by analyzing the function $\beta_i(\eta)$ for a minimum, since $\beta_i = H_{c_2}(i)/H_0$.

At the same time the equilibrium value of η_0 for which β_i reaches a minimum is determined. Anticipating a little, we may say that η_0 is the dimensionless distance from the boundary with the vacuum to the point at which a maximum density of the superconducting electrons is reached in the superconducting surface layer.

Let us carry out this program.

Derivation and Solution of the Principal

Equation

The general solution of equation (6) has the form

$$\Psi_1 = e^{-\xi^2/2}\left[C_1\Phi\left(\frac{1-\beta_1}{4}, \frac{1}{2}; \xi^2\right) + C_2\xi\Phi\left(\frac{3-\beta_1}{4}, \frac{3}{2}; \xi^2\right)\right], \tag{9}$$

$$\Psi_2 = e^{-\xi^2/2}\left[C_3\Phi\left(\frac{1-\beta_2}{4}, \frac{1}{2}; \xi^2\right) + C_4\xi\Phi\left(\frac{3-\beta_2}{4}, \frac{3}{2}; \xi^2\right)\right]. \tag{10}$$

Here $C_1, \ldots, C_4$ are arbitrary constants and Φ is a degenerate hypergeometric function.

For brevity of writing we introduce the following notation

$$\alpha \equiv \frac{1-\beta_1}{4} \qquad\qquad \gamma \equiv \frac{1-\beta_2}{4}, \tag{11}$$

$$F_\alpha(\xi) \equiv \Phi\left(\frac{1-\beta_1}{4}, \frac{1}{2}; \xi^2\right), \quad F_\gamma(\xi) \equiv \Phi\left(\frac{1-\beta_2}{4}, \frac{1}{2}; \xi^2\right), \tag{12}$$

$$\Phi_\alpha(\xi) \equiv \Phi\left(\frac{3-\beta_1}{4}, \frac{1}{2}; \xi^2\right), \quad \Phi_\gamma(\xi) \equiv \Phi\left(\frac{3-\beta_2}{4}, \frac{3}{2}; \xi^2\right), \tag{13}$$

$$\xi_0 \equiv -\eta, \quad \xi_1 \equiv s - \eta. \tag{14}$$

Then solutions (9) and (10) may be written

$$\Psi_1 = e^{-\xi^2/2}[C_1 F_\alpha(\xi) + C_2\xi\Phi_\alpha(\xi)], \tag{15}$$

$$\Psi_2 = e^{-\xi^2/2}[C_3 F_\gamma(\xi) + C_4\xi\Phi_\gamma(\xi)]. \tag{16}$$

One of the coefficients in equation (16) may be eliminated by using the boundary condition (8) $\Psi_2/\xi_{\to\infty} = 0$ and the asymptotic of the degenerate hypergeometric function $\Phi(u, v; x) \sim$ $\sim (\Gamma(v)/\Gamma(u)) e^x x^{u-v}$. We use the following relation between C_3 and C_4:

$$C_3 = -\frac{1}{2}\frac{\Gamma(\gamma)}{\Gamma(\gamma + 1/2)} C_4. \tag{17}$$

Now we may put solution (17) in the form

$$\Psi_2 = Ce^{-\xi^2/2}\Psi_\gamma(\xi), \tag{18}$$

where C is an arbitrary constant

$$\Psi_\gamma = -\frac{1}{\Gamma(2\gamma + 1)} F_\gamma(\xi) + \psi_\gamma\xi\Phi_\gamma(\xi), \tag{19}$$

$$\psi_\gamma = 2\sqrt{\pi}e^{-p\gamma}\gamma/\Gamma^2(\gamma + 1), \quad p = 2\ln 2. \tag{20}$$

If we now write down the three remaining unused conditions (8), allowing for the notation (14) for Ψ_1 (15) and Ψ_2 (18), and eliminate C_1, C_2, and C, we obtain the following principal equation:

$$[\sqrt{k}F_\alpha(\xi_1)\Psi'_\gamma(\xi_1) - \sqrt{k}\xi_1 F_\alpha(\xi_1)\Psi_\gamma(\xi_1) -$$

$$-\frac{1}{\sqrt{k}}F'_\alpha(\xi_1)\Psi_\gamma(\xi_1) + \frac{1}{\sqrt{k}}\xi_1 F_\alpha(\xi_1)\Psi_\gamma(\xi_1)][\xi_0^2\Phi_\alpha(\xi_0) - \Phi_\alpha(\xi_0) - \xi_0\Phi'_\alpha(\xi_0)] =$$

$$= \left[\frac{1}{\sqrt{k}}\Phi_\alpha(\xi_1)\Psi_\gamma(\xi_1) + \frac{1}{\sqrt{k}}\xi_1\Phi'_\alpha(\xi_1)\Psi_\gamma(\xi_1) - \frac{1}{\sqrt{k}}\xi_1^2\Phi_\alpha(\xi_1)\Psi_\gamma(\xi_1) -\right.$$

$$\left. -\sqrt{k}\xi_1\Phi_\alpha(\xi_1)\Psi'_\gamma(\xi_1) + \sqrt{k}\xi_1^2\Phi_\alpha(\xi_1)\Psi_\gamma(\xi_1)\right][F'_\alpha(\xi_0) - \xi_0 F_\alpha(\xi_0)]. \tag{21}$$

The prime here denotes the derivative of the corresponding function with respect to ξ at the points ξ_1 and ξ_0 respectively.

This equation will constitute the foundation of the following treatment; it includes the relation between β_i and η mentioned earlier in implicit form.

Our problem now reduces to obtaining this relation explicitly. In the general case, as inspection of Eq. (21) readily confirms, this problem cannot be solved. However, a complete solution may be secured by the method of successive approximations for the two limiting cases $s \ll 1$ and $s \gg 1$. Analysis of these two limiting cases will also provide a qualitative answer for intermediate values of s. For both limiting cases we assume that η is a small parameter. As we shall see later, this assumption is fully justified.

Case in Which s ≪ 1. In this case ξ_1 and ξ_0 are small parameters. The functions F_α, Φ_α, Ψ_γ and their derivatives are expanded in series in powers of ξ_0 and ξ_1. We seek α and γ in the form of series in powers of ξ_0 and ξ_1:

$$\alpha = \alpha_0 + \alpha_1 + \alpha_2 + \cdots,$$
$$\gamma = \gamma_0 + \gamma_1 + \gamma_2 + \cdots.$$

The index indicates the order of smallness of the corresponding term. Substituting these series into the fundamental equation (21) and separating terms of the same order of smallness, we may calculate α_n and γ_n, where n = 0, 1, 2,... .

Carrying the calculations out to the third order, we have

$$\gamma = \frac{1}{2\sqrt{\pi}}\,\eta + \frac{1}{4\pi}(p-4)\,\eta^2 + f\eta^3 + \frac{1}{2\sqrt{\pi}}\frac{k-1}{k}(s\eta^2 - \eta s^2) + \frac{k^2-1}{6\sqrt{\pi}k}\,s^3, \qquad (22)$$

where $f = \frac{1}{8\pi\sqrt{\pi}}\left(2d_2 - C^2 - 12p + \frac{3}{2}p^2 + 16\right) \cong 0.0134$. Here C ~ 0.577 is Euler's constant, $d_2 \cong \frac{1}{2}(C^2 - \zeta(2)) \cong 0.656$; $\zeta(x)$ is a Riemann zeta function, and p = 2 ln 2.

It was mentioned earlier that the relationship $\beta_2(\eta)$ had to be examined for a minimum; however, $\gamma = (1-\beta_2)/4$, and this is thus equivalent to examining $\gamma(\eta)$ for a maximum. Until now η has been a free parameter. The η_0 which corresponds to the minimum of β_2 (maximum of γ) will also correspond to the minimum free energy of the system [8]; hence it is this value which will be realized under equilibrium conditions.

Let us find η_0:

$$\frac{\partial\gamma}{\partial\eta} = \frac{1}{2\sqrt{\pi}} + \frac{1}{2\pi}(p-4)\,\eta_0 + 3f\eta_0^2 + \frac{1}{\sqrt{\pi}}\frac{k-1}{k}s\eta_0 - \frac{1}{2\sqrt{\pi}}\frac{k-1}{k}s^2 = 0. \qquad (23)$$

The solution of this quadratic equation in η_0 will be

$$\eta_0 = a\left(1 - b\frac{k-1}{k}s\right) - \sqrt{a^2\left(1 - b\frac{k-1}{k}s\right)^2 - c\left(1 - \frac{k-1}{k}s^2\right)}, \qquad (24)$$

where

$$a = \frac{4-p}{12\pi f} \cong 5.16; \quad b = \frac{2\sqrt{\pi}}{4-p} \cong 1.36, \quad c = \frac{1}{6\sqrt{\pi}f} \cong 7.01.$$

The minus sign in front of the root is taken so that in the limiting case k →1 the value of η_0 will tend toward $\eta_0 = a - \sqrt{a^2 - c} \cong 0.73$, which gives $H_{c_3}/H_{c_2} = 1.67$, very close to the value of 1.69 obtained in [1].

After substituting the resultant value of η_0 in equation (22) and expressing γ in terms of β_2 and β_2 in terms of H_0, we find the maximum possible H_0, i.e., H_{c_3}, as a function of the parameters of the material $k = \varkappa_1/\varkappa_2$ and the film thickness s. However, complicated expressions are involved and these are difficult to analyze. Since the whole consideration has been based on the smallness of s and η, we may analyze the case $s \ll 1$ and confine attention to terms linear in s. Making the necessary arithmetical calculations, we obtain

$$\eta_0 \cong 0.73 + 1.2 \frac{k-1}{k} s, \ \gamma(\eta_0) = 0.100 + 0.15 \frac{k-1}{k} s. \tag{25}$$

Thus we obtain a maximum value for γ. The corresponding minimum value of β_2 will according to (11) and (7) be

$$\beta_{2\,\mathrm{min}} = 1 - 4\gamma_{\mathrm{max}} = \frac{H_{c_2}(2)}{H_{c_3}}.$$

From this we immediately obtain the desired expression for H_{c_3}

$$\frac{H_{c_3}}{H_{c_2}(2)} = 1.67\left(1 + \frac{k-1}{k}s\right). \tag{26}$$

In ordinary units this formula reads

$$\frac{H_{c_3}}{H_{c_2}(2)} = 1.67\left(1 + \frac{\varkappa_1 - \varkappa_2}{\varkappa_1}d\sqrt{\frac{3.4H_{c_2}(2)e}{\hbar c}}\right). \tag{27}$$

Formula (27) may be written in a different way. Since $\varkappa_2 = \frac{2\sqrt{2}e}{\hbar c}\delta_{02}^2 H_{\mathrm{KM}}$ and $H_{c_2}(2) = \sqrt{2}\,\varkappa_2 H_{\mathrm{KM}}$, we have $H_{c_2}(2) = \varkappa_2^2 \frac{\hbar c}{2e\delta_{02}^2}$ (here δ_0 is the depth of penetration). Substituting this value of $H_{c_2}(2)$ into formula (27), we obtain

$$\frac{H_{c_3}}{H_{c_2}(2)} = 1.67\left(1 + \frac{\varkappa_1 - \varkappa_2}{\varkappa_1}\sqrt{1.7}\varkappa_2\frac{d}{\delta_{02}}\right). \tag{28}$$

We see from the result that for $\varkappa_1 \to \varkappa_2$ or $d \to 0$ formula (28) transforms into $H_{c_3} = 1.67\,H_{c_2}$, which differs little from the $H_{c_3} = 1.69\,H_{c_2}$, of Saint-James and Gennes. This difference occurs because we solved Eq. (21) approximately, confining attention to the third order. In the fourth-order approximation H_{c_3}/H_{c_2} equals 1.68 for a clean surface.

Case in Which $s \gg 1$. In this limiting case $\xi_0 = -\eta$ will be considered as a small parameter. This is justified because in the limit $s \to \infty$ the equilibrium value of η tends to 0.73. The value of $\xi_1 = s - \eta$, however, will be regarded as large ($\xi_1 \gg 1$).

Let us now turn to the original, very general equation (21). For the case under consideration this may be more conveniently written in another way, in terms of parabolic cylindrical functions:

$$\left[-\sqrt{k}\frac{dD_{-2\gamma}(\sqrt{2}\xi_1)}{d\xi}D_{-2\alpha}(-\sqrt{2}\xi_1) + \frac{1}{\sqrt{k}}D_{-2\gamma}(\sqrt{2}\xi_1)\frac{dD_{-2\alpha}(-\sqrt{2}\xi_1)}{d\xi}\right]\times$$

$$\times\left[\frac{2\Gamma(\alpha+\frac{1}{2})}{\Gamma(\alpha)}A_\alpha(\xi_0) + B_\alpha(\xi_0)\right] = \left[\sqrt{k}\frac{dD_{-2\gamma}(\sqrt{2}\xi_1)}{d\xi}D_{-2\alpha}(\sqrt{2}\xi_1) - \right.$$

$$\left. - \frac{1}{\sqrt{k}}D_{-2\gamma}(\sqrt{2}\xi_1)\frac{dD_{-2\alpha}(\sqrt{2}\xi_1)}{d\xi}\right]\left[\frac{2\Gamma(\alpha+\frac{1}{2})}{\Gamma(\alpha)}A_\alpha(\xi_0) - B_\alpha(\xi_0)\right]. \tag{29}$$

Here we have used the notation

$$A_\alpha(\xi_0) = \xi_0^2 \Phi_\alpha(\xi_0) - \Phi_\alpha(\xi_0) - \xi_0 \Phi_\alpha'(\xi_0),$$

$$B_\alpha(\xi_0) \equiv F_\alpha'(\xi_0) - \xi_0 F_\alpha(\xi_0).$$

Since $\xi_1 \gg 1$, we may use the asymptotic expansion of parabolic cylindrical functions

$$\begin{aligned}
D_p(z) &\sim e^{-z^2/4} z^p \quad (z > 0), \\
D_p(z) &\sim \frac{\sqrt{2\pi}}{\Gamma(-p)} e^{z^2/4} |z|^{-p-1} \quad (z < 0).
\end{aligned} \tag{30}$$

After making some reasonably simple transformations we now obtain the following asymptotic expressions for the original Eq. (29):

$$\frac{2\alpha}{\Gamma(2\alpha+1)} \psi_\alpha A_\alpha(\xi_0) + \frac{2\alpha}{\Gamma^2(2\alpha+1)} B_\alpha(\xi_0) + \omega = 0, \tag{31}$$

$$\omega = \frac{1}{\sqrt{\pi}} e^{-p\alpha} \xi_1^{1-4\alpha} e^{-\xi_1^2} \frac{k-1}{k+1} \left[\psi_\alpha A_\alpha(\xi_0) - \frac{1}{\Gamma(2\alpha+1)} B_\alpha(\xi_0) \right], \tag{32}$$

where p = 2 ln 2 and ψ_α is given by Eq. (20).

Expression (31) may be considered as an implicit form of expressing the functional relationship $\alpha = \alpha(\xi_0, \xi_1)$. Let us find this relationship. It is easy to see that for $\xi_1 \to \infty$ the value of $\omega \to 0$. We already know the solution of (31) without the term ω. This corresponds to the simple case of the clean surface of a semiinfinite superconducting space [1], and in our method corresponds to the case k = 1 in the previous paragraph. It is thus natural to solve equation (31) by the method of successive approximations.

Let use seek α in the form

$$\alpha = \varepsilon_0(\xi_1) + (b_1 + \varepsilon_1(\xi_1))\xi_0 + (b_2 + \varepsilon_2(\xi_1))\xi_0^2 + \ldots. \tag{33}$$

The numbers $b_1, b_2, \ldots$ are known from previous consideration ($b_1 = -\frac{1}{2}\sqrt{\pi}$, $b_2 = (p-4)/4\pi$) since for $\xi_1 \to \infty$ the quantities $\varepsilon_0, \varepsilon_1, \varepsilon_2 \to 0$ and α depends only on ξ_0 by virtue of the law already derived. Let us confine attention to the terms linear in all the ε functions. We expand (31) in series in powers of ξ_0 by substituting (33) into the first two terms and the expression $\alpha = b_1\xi_0 + b_2\xi_0^2$ into the term ω. Comparing coefficients for the same powers of ξ_0 we obtain finally:

$$\varepsilon_0 = -\frac{1}{\sqrt{\pi}} \frac{k-1}{k+1} \xi_1 e^{-\xi_1^2},$$

$$\varepsilon_1 = -\frac{1}{\pi} \frac{k-1}{k+1} \xi_1 (2 + C + 2\ln\xi_1) e^{-\xi_1^2},$$

$$\varepsilon_2 = \frac{1}{2\pi\sqrt{\pi}} \frac{k-1}{k+1} \xi_1 \left[pC - 8C - C^2 - \frac{p^2}{4} + 4p - 8 + (2p - 4C - 16)\ln\xi_1 - 4\ln^2\xi_1 \right] e^{-\xi_1^2},$$

where p = 2 ln 2, C $\cong$ 0.577 is Euler's constant.

In order to determine the critical field H_{c_3}, as before, we have to find the value of η_0, which corresponds to the maximum value of $\alpha = \varepsilon_0 - (b_1 + \varepsilon_1)\eta + (b_2 + \varepsilon_2)\eta^2$. The latter expression follows from (33), since $\xi_0 = -\eta$.

Let us seek η_0:

$$\frac{d\alpha}{d\eta} = -(b_1 + \varepsilon_1) + 2(b_2 + \varepsilon_2)\eta_0 = 0.$$

Finding η_0 from this and substituting into the expression for α, we have

$$\alpha_{\max} = -\frac{b_1^2}{4b_2}\left(1 + 2\frac{\varepsilon_1}{b_1} - \frac{\varepsilon_2}{b_2}\right) + \varepsilon_0.$$

Substituting all the necessary numerical values and carrying out the arithmetical calculations, we finally obtain

$$\frac{H_{c_3}}{H_{c_2}} = \frac{1}{\beta_{\min}} = \frac{1}{1 - 4\alpha_{\max}} = 1.62\left[1 - \frac{1}{\sqrt{\pi}}\frac{k-1}{k+1}\xi_1(3.43 + 2.41\ln\xi_1 + 1.90\ln^2\xi_1)e^{-\xi_1^2}\right]. \tag{34}$$

We see from this expression that for $\varepsilon_1 \to \infty$, $H_{c_3}/H_{c_2} \to 1.62$. This value differs from the 1.67 obtained by formula (28) simply because in formula (28) we considered the third order, while here we have confined attention to the second order in the expansion of α in powers of ξ_0. This is quite unimportant, since we are interested in the correction term.

Let us put $\xi_1 \cong s$ in expression (34) and convert to ordinary units:

$$H_{c_3}(d) = H_{c_3}(\infty)\left\{1 - \frac{1}{\sqrt{\pi}}\frac{\varkappa_1 - \varkappa_2}{\varkappa_1 + \varkappa_2}\frac{\varkappa_1 d}{\delta_{01}}\left[4.5 + 3.1\ln\left(\sqrt{1.7}\,\frac{\varkappa_1 d}{\delta_{01}}\right) + 2.5\ln^2\left(\sqrt{1.7}\,\frac{\varkappa_1 d}{\delta_{01}}\right)\right]e^{-1.7(\varkappa_1 d/\delta_{01})^2}\right\}. \tag{35}$$

Discussion of Results

Comparison of formulas (28) and (35) first of all reveals the monotonic character of the variation in H_{c_3} with film thickness. This relation is shown schematically in Fig. 1 (on the assumption that $\varkappa_1 > \varkappa_2$). Thus we see quite strictly, without making any further assumptions, that the fall in the electron mean free path l in the surface layer of the superconductor leads to a rise in the third critical field. Such a reduction in mean free path always occurs after mechanical treatment (machining) of the surface. Hence the facts indicated at the very beginning of this article receive a qualitative explanation. Let us turn, for example, to [3].

Let us suppose that the electropolishing of the In-Pb surface has led to a considerable change in the electron mean free path in the surface layer of the sample. This should lead to a rise in H_{c_3}/H_{c_2} above 1.7. Etching the surface reduces the thickness of the deformed layer and correspondingly reduces H_{c_3}/H_{c_2}. Finally, prolonged annealing evens out the characteristics of the surface layer and the material as a whole, the layer as such disappears, and H_{c_3}/H_{c_2} becomes equal to 1.7.

It is interesting to note that the calculation carried out also embraces the case of $\varkappa_1 < \varkappa_2$. In this case we should expect a fall in H_{c_3}/H_{c_2} as compared with 1.7.

Let us direct our attention to another aspect of the study of surface superconductivity. It is well known that a drawn wire has a fibrous microstructure, the fibers being drawn-out grains. In the recent paper [9] it was noticed that the critical field of a wire made from a superconductor of the second kind rose above H_{c_2}. The authors associated this with the appearance of surface conductivity inside the wire along the boundaries of the grains drawn out in the drawing process. Here it must immediately be stipulated that a substantial difference of H_{c_3} from H_{c_2} should only be expected when the superconductor borders a dielectric, and not a metal [7]. Such a situation may arise if, for example, the grain boundaries are covered with oxides. On the other

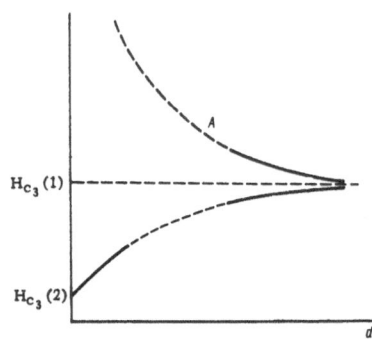

Fig. 1. Dependence of H_{c_3} on the film thickness for a system comprising a film deposited on the plain surface of a massive superconductor. For comparison the figure shows an analogous relation (curve A) for a film on dielectric substrate (Abrikosov [8]).

hand, the grain boundaries are enriched with impurities, dislocations, etc., which leads to a reduction in the mean free path of an electron near the surface of the grain. According to our calculation, this increases H_{c_3} and hence the critical field of the wire.

LITERATURE CITED

1. D. Saint-James and P. G. Gennes, Phys. Lett., 7:306 (1963).
2. C. F. Hempstead and J. B. Kim, Phys. Rev. Lett., 12:145 (1964).
3. S. Gygax, J. L. Olsen, and R. H. Kropschot, Phys. Lett., 8:228 (1964).
4. P. R. Doidge and Kwan Sik-Hung, Phys. Lett., 12:82 (1964).
5. L. P. Gor'kov, Zh. Éksp. Teor. Fiz., 37:1407 (1959).
6. V. L. Ginzburg and L. D. Landau, Zh. Éksp. Teor. Fiz., 20:1064 (1950).
7. R. O. Zaitsev, Zh. Éksp. Teor. Fiz., 50:1055 (1966).
8. A. A. Abrikosov, Zh. Éksp. Teor. Fiz., 47:720 (1964).
9. S. J. Williamson and J. K. Furdyana, Phys. Lett., 21:376 (1966).

CRITICAL MAGNETIC FIELDS OF
HIGH-FIELD SUPERCONDUCTORS

B. G. Lazarev, O. N. Ovcharenko, and A. A. Matsakova

The critical magnetic fields H_{K_2} and the field corresponding to the initial penetration of the magnetic flux H_{K_1} are measured for the intermetallic compounds Nb_3Sn and V_3Ga and worked Nb–Ti and Nb–Zr–Ti alloys. The relation between H_{K_1} and the state of the alloys is studied. The field corresponding to the penetration of magnetic flux increases with increasing inhomogeneity of the sample and equals about $10^{-2}H_{K_2}$.

It is well known that superconductors of the second kind are characterized by the existence of two critical magnetic fields, namely [1, 2]: H_{K_1}, the magnetic field at which the magnetic flux starts penetrating into the superconductor (the electrical resistance of the sample still remains zero) and H_{K_2}, the magnetic field at which the electrical resistance of the material is completely restored (the material passes from the superconducting to the normal state). Homogeneous (single-phase) alloys of the solid-solution type belong to this class of superconductors [1].

Practically valuable superconductors with high critical parameters (superconducting transformation temperature, critical current) also have two critical magnetic fields: the field at which the magnetic flux starts penetrating and the field corresponding to the complete restoration of electrical resistance. However, these superconductors (alloys of transition metals and intermetallic compounds) are not homogeneous. For example, deformable superconducting alloys are characterized by the precipitation of equilibrium phases forming a fine network of superconducting paths [3]. In the intermetallic compounds Nb_3Sn, V_3Ga, etc. there is a polymorphic transformation at a temperature above critical [4, 5]. The conclusions of the modern theory of superconducting alloys are not applicable to such superconductors. The critical magnetic fields for these superconducting materials are nevertheless arbitrarily called H_{K_1} and H_{K_2} as in the case of superconductors of the second kind such as Bi–Pb and Pb–Tl alloys [1].

The lower critical magnetic fields H_{K_1} have been measured for the superconducting ternary alloys Nb–Zr–Ti over a wide range of concentrations and also for the intermetallic compounds Nb_3Sn and V_3Ga.

The values of H_{K_1} were determined from measurements of magnetic induction and magnetic moment as functions of the external magnetic field.

The magnetic flux was measured by the ordinary ballistic method. The magnetic field was created in a solenoid with a superconducting winding developed and made in the Physicotechnical Institute of the Academy of Sciences of the Ukrainian SSR. The solenoid constant was 2425 Oe/A. The inhomogeneity of the magnetic field along the axis of the solenoid at a distance of ± 1.5 cm from the center was 1%.

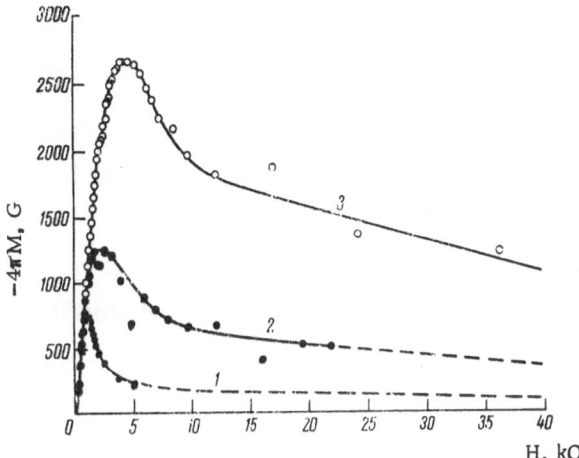

Fig 1. Dependence of the magnetic moment on the external magnetic field for the ternary alloy Nb–22%Zr–30%Ti. 1) Cast, 2) worked, 3) annealed after working.

The measurement was carried out at a temperature of 4.2°K in magnetic fields up to 40 kOe.

Figure 1 shows the magnetic moment as a function of the external magnetic field for the ternary alloy Nb–22at.%Zr–30at.%Ti in three states: cast (quenched), worked (into a wire), and annealed after working. We see from this figure that on raising the magnetic field from zero to a certain value (H_{K_1}) the magnetic flux rises linearly, i.e., the Meissner effect is realized; the sample is completely diamagnetic, the magnetic flux not penetrating into the superconductor.

On exceeding the value of H_{K_1} the magnetic moment starts deviating from the linear relationship; this indicates the penetration of the magnetic flux into the superconductor.

For a sample in the cast state H_{K_1} = 600 Oe, in the worked state 850 Oe, and in the annealed state after working 1500 Oe. The maximum value of the magnetic moment is realized for the quenched alloy in a magnetic field of 1000 Oe, for the worked alloy in a field of 2500 Oe, and for the annealed alloy in a field of 4500 Oe.

The value of the magnetic moment is different in all three samples and rises as the structure of the precipitates leading to the increase in critical current becomes more complex. It follows from measurements of critical current (at 4.2°K in a magnetic field of 20 kOe) that the critical current density in the cast sample is 10^3 A/cm^2, in the worked sample (wire) 10^4 A/cm^2, and in the sample annealed after working 10^5 A/cm^2.

Thus the state of the sample with the greatest critical current corresponds to the greatest magnetic moment.

The upper critical magnetic field H_{K_2} measured in pulsed magnetic fields* was the same for all three samples (78 kOe).

For ternary superconduction alloys in the cast state the value of H_{K_1} depends very slightly on the composition, and for a wide range of concentrations between Nb–50Zr–15Ti (at.%) and Nb–5Zr–48Ti (at.%) H_{K_1} = 500-600 Oe.

The values of H_{K_1} have also been measured for the intermetallic compounds Nb$_3$Sb (made by the diffusion method [6]) and V$_3$Ga (obtained by fusion of the components in the Institute of Metal Physics of the Academy of Sciences of the Ukrainian SSR). For Nb$_3$Sn H_{K_1} = 500 Oe, which is close to published values [7, 8]; for V$_3$Ga H_{K_1} = 1500 Oe, which differs considerably from the values published in [7]. Evidently the conditions of obtaining the Nb$_3$Sn were more favorable for producing alloys in an equilibrium state than in the case of V$_3$Ga.

It is interesting to note that in alloys and compounds with a large value of H_{K_2} (about 100 kOe) H_{K_1} is as small as in the case of homogeneous alloys of the Pb–Bi type [1]. The value of H_{K_1} for superconducting alloys and compounds with a high value of the upper critical magnetic field is about 0.01 H_{K_2}.

*See the article by B. G. Lazarev et al., "Superconducting Properties of Niobium-Base Alloys," this volume, p. 86.

LITERATURE CITED

1.   L. V. Shubnikov, V. I. Khotkevich, Yu. D. Shepelev, and Yu. N. Ryabinin, Zh. Éksp. Teor. Fiz., 7:221 (1937).
2.   A. Abrikosov, Zh. Éksp. Teor. Fiz., 32:1442 (1957).
3.   B. G. Lazarev, V. K. Khorenko, L. A. Kornienko, A. I. Krivok, A. A. Matsakova, and O. N. Ovcharenko, Zh. Éksp. Teor. Fiz., 45:2068 (1963).
4.   B. W. Batterman and C. S. Barrett, Phys. Rev. Lett., 13:390 (1964).
5.   M. J. Goringe and U. Valdre, Phys. Rev. Lett., 14:823 (1965).
6.   V. S. Kogan, A. I. Krivko, B. G. Lazarev, et al., Collection "Metallography and Metallo-physics of Superconductors" [in Russian], Izd. "Nauka" (1965), p. 76.
7.   P. S. Schwartz, Phys. Rev. Lett., 9:448 (1965).
8.   R. Hecht, RCA Review, 25:453 (1964).

REASONS FOR CHANGES IN
THE SUPERCONDUCTING PROPERTIES OF Nb₃Sn

V. N. Svechnikov, V. M. Pan, and Yu. I. Beletskii

A review of investigations relating to the phase diagrams of the Nb–Sn system and the superconducting properties of the compound Nb_3Sn is presented. The high-niobium side of the Nb–Sn phase diagram is studied by x-ray structural and microstructural analysis. The compound Nb_3Sn has a range of homogeneity and is stable right down to room temperature. Measurements of T_K show that the critical temperature of Nb_3Sn increases linearly with the amount of tin present.

The first attempt at constructing a phase diagram for the Nb–Sn system was made by Agafonova et al. [1] after it had first been shown in [2, 3] that the compound Nb_3Sn, having a crystal structure of the β-W type (A15), was a superconductor with a superconducting-transformation temperature of 18.05 ± 0.1°K.

It was later shown in [4] that the critical magnetic field for Nb_3Sn was unusually high, 70 kOe at 4.2°K.* This stimulated further investigations into the superconducting properties of Nb_3Sn as a promising material for superconducting solenoids [5-11]. Despite the brittleness of Nb_3Sn, methods of obtaining wire from this compound were developed [5, 10, 11]. It was shown in [12-16] that the superconducting properties of Nb_3Sn wire depended sharply on the conditions of heat treatment. On raising the annealing temperature to 850-860°C [14], 900°C [12-13], 930-950°C [15] or 1000°C [16] the critical temperature and critical current density rose sharply. On further raising the annealing temperature these parameters diminished. This phenomenon cannot be explained from the point of view of the diagram plotted in [1]. Subsequent metallographical and physical research work was undertaken in two directions: firstly in order to refine, supplement, and correct the phase diagram as a whole, and secondly in order to discover the physical nature of the change in superconducting properties. New forms of Nb–Sn phase diagram were proposed in [14, 17-23]; these differ from the earlier one [1] in that they indicate the existence of other compounds as well as Nb_3Sn in the Nb–Sn system and refute the occurrence of phase separation in the liquid state and the monotectic reaction. In addition to this, individual regions of the Nb–Sn system and new compounds within the latter were studied in [15, 24-36, 43].

On the basis of the results given in these various papers, Table 1 presents all the proposed compounds in the Nb–Sn system, their chemical compositions, and their temperature ranges of stability.

Thus the majority of authors consider that niobium may form three compounds with tin. The disagreement mainly concerns the composition of the compounds and the temperature

---

* In later investigations [9, 44] a critical field of the order of 200 kOe at 4.2°K was found.

TABLE 1. Proposed Compounds in the Nb–Sn System According to Various Research Workers (Numerators give temperatures (°C) for stability, denominators give concentrations (at.% Nb) for stability)

	Nb_6Sn	$⁻Nb_6Sn$	Nb_6Sn	Nb_6Sn_3	Nb_6Sn_3	Nb_6Sn_3	Nb_2Sn_3	$NbSn_3$
Savitskii et al. [1]	Room—2050 / 20.8—22.6	Room—2000 / 75						
Lazarev et al. [14]		Room—2000 / 75						
Kolbe, Rosner [15]		—		Room—850 / —	—		—	
Wyman et al. [17]		Room—730 / 25.1—26.8	Room—690 / 30.2					
Reed et al. [18]		860—2000			775—890 / 60		Room—863 / 39.0—41.3; Room—850 / 40	
Enstrom et al. [19]		775—2000 / 75—80			600—1175 / 56.1		Room—1050 / 38.6—40	
Enstrom ct al. [20]		775—2000 / 75—79			600—925 / 54.5—56.1; Below 915 / 59			
Schadler et al. [21]		78.5						
Ellis, Wilhelm [22]		Room—2130			Room—915	NbSn Room—850 / 50	Room—820	
Levinstein et al. [23]		805—2000			805—925			Room—850 / 35—40; Below 840 / 33
Enstrom et al. [24]		75			56; 775—875		38; Room—875	Room—950
Reed, Gatos [25]		—			44 ± 2			
VanVucht et al. [26, 30, 31]		—						
Lazarev et al.		Room—2000 / —						— / 33.3

ranges of stability. However, in the majority of the later investigations [20-23, 26, 30, 31, 36, 43] there is a considerable convergence of views on the composition of compounds containing more tin than Nb_3Sn. The compound richest in tin is given the formula $NbSn_2$. The crystal structure of this compound is given in a number of papers as orthorhombic, of the $CuMg_2$ type, with the following lattice constants (in Å):

a	5.63	5.64	5.645	5.6477	5.655
b	9.85	9.86	9.852	9.860	9.860
c	18.96	19.13	19.126	19.127	19.152
References	[23]	[20]	[26]	[31]	[22]

The compound lying between Nb_3Sn and $NbSn_2$ is called Nb_3Sn_2 by the majority of authors. In later papers [20, 22, 30, 37, 43] it was found that the crystal structure of this compound was orthorhombic, evidently of the $\beta\text{-}Ti_6Sn_5$ type, with the following parameters (in Å):

a	5.65	5.6549	5.656
b	9.21	9.2037	9.199
c	16.84	16.814	16.843
References	[20]	[30,37]	[22, 43]

Considering the structural type (which is associated with the formula A_6B_5) and also the fact that in [19, 20, 24] the chemical composition of the compound approximately given as Nb_3Sn_2 by the authors was in fact Nb_6Sn_5 (Table 1), we tend to prefer the latter formula.

The authors of several papers consider that Nb_3Sn is only stable above 775° [19, 20], 805° [23], or 860° [18], while Nb_6Sn_5 (as we shall now call this compound despite its usual name of Nb_3Sn_2) is stable above 600° [19, 20] 775° [18, 25] or 805° [23]. The authors of another group of papers indicate that Nb_3Sn is stable down to room temperature [1, 14, 22] like Nb_6Sn_5 [22].

It should be noted that in those papers in which only $NbSn_2$ is regarded as stable at low temperatures the samples were prepared by sintering powder mixtures of niobium and tin (maximum annealing period 300 h at 750°C [23]). The results obtained by the method of diffusion couples lead different authors to different conclusions. In [14, 22] it was concluded that Nb_3Sn was formed at temperatures at least up to 600°C, while in [20, 23] it was considered that there was no compound Nb_3Sn in the diffusion layers at temperatures below 775° [19] or 805° [23].

If the hypothesis as to the stability of Nb_3Sn and Nb_6Sn_5 down to room temperature is correct, we must conclude that the rate of diffusion of the tin atoms in niobium and in the compounds is extremely low.

In ending our review of papers treating the Nb–Sn diagram in various ways, we note certain properties of the known phases in this system.

Alpha Solid Solution Based on Niobium. According to the results of [1] the solubility of tin varies from 10 wt.% at 800 to 1400° to 14% at 2000°C. The lattice constant meanwhile changes very little. It was found in [22] that the solubility of tin in niobium varied from 2.5 wt.% at 550° to 3.0% at 1000°C. According to [19, 20] the solubility of tin in niobium at 700 to 1000° is 6% and at 2000° around 12 at.%. The lattice parameter varies from 3.312 to 3.329 Å.

Compound Nb_3Sn. In the majority of published papers the concentration range of homogeneity of the compound Nb_3Sn is regarded as very narrow and is shown on the diagram as a vertical line [1, 14, 17, 18, 23]. In [18-20, 32, 33] there are indications that the composition of Nb_3Sn may differ from stoichiometric in the direction of niobium. It is also indicated in [19, 20] that the lattice constant of Nb_3Sn in the approximate concentration range 75 to 79 at.% Nb varies from 5.280 to 5.289 Å at high temperatures (1300 to 1400°C); according to [32] it varies from 5.2803 to 5.291 Å (between 81.5 and 75 at.% Nb), according to [33] from 5.282 to 5.290 Å (between 82.3 and 75.1 at.% Nb), according to [34] from 5.2826 to 5.2887 Å (between 78.5 and 72.8 at.% Nb). This disagrees with the results of [38], according to which the lattice constant of Nb_3Sn remained constant (a = 5.291 ± 0.002 Å) between 50 and 90 at.% Nb.

There are also contradictions in the determination of the superconducting properties of Nb_3Sn as functions of composition. It was found in [18, 38] that the superconducting transformation temperature rose slightly on increasing the niobium concentration and reached 18.5°K [18] or 18.19°K [38] at 80 at.% Nb. In [11, 29, 32, 33, 40] it was found, however, that the highest critical current density and critical temperature belonged to samples with a greater tin content, while samples with a composition close to 80 at.% Nb had a low current density and T_K = 5.6°K [29], 6.2°K [40], 7°K [32, 33] and 9.0°K [11].

Summarizing all the foregoing we come to the conclusion that the results obtained in numerous recent investigations into phase equilibria in the Nb—Sn system fail to explain the worsening in the superconducting properties on raising the temperature of heat treatment above 900 to 1000°C.

Various hypotheses regarding the instability of the superconducting properties have recently been advanced [11, 16, 19, 39, 40]. An attempt was made in [16] to explain the reduction in the critical current density on raising the temperature of heat treatment from a dislocation point of view: The critical current density depends on the dislocation density, and on raising the annealing temperature above 900 to 1000°C the dislocation density in the Nb_3Sn grains falls sharply. It was found in [11] that Nb_3Sn may occur in the disordered state, a certain number of niobium atoms being situated in the tin lattice points and vice versa. Certain authors have tried to associate the transformation into the disordered state with the instability of the superconducting properties; however, it was shown in [19] that no order/disorder transformation occurred up to 1400°C.

The authors of [39, 40] came to the conclusion that the high parameters of the superconducting compound Nb_3Sn (and probably other compounds with the β-W type of structure) were associated with the existence of close-packed rows (chains) of niobium atoms. The reduction in the transformation temperature and other superconductivity parameters occurs when these chains are broken. It was found that the evaporation of the tin during high-temperature vacuum annealing and also any deviations from the stoichiometric composition of Nb_3Sn in the niobium direction led to a reduction in the lattice parameter of Nb_3Sn and to a fall in the transformation temperature and critical field. The authors explained this by the formation of vacancies at the points occupied by the tin atoms; after reaching a certain critical vacancy concentration these jumped into the niobium lattice points, breaking the close-packed rows of niobium atoms.

We should also mention another paper [41], in which it was found that Nb_3Sn, like other compounds with the β-W structure, underwent a phase transformation of the martensite type at 36.0°K, with the formation, as it would appear, of a tetragonal lattice, as in the case of V_3Si [42]. There is an indication that this phase transformation may be affected by various factors (in particular the conditions of formation of the Nb_3Sn), thus changing the superconducting properties.

It was because of the existing disagreement in relation to the Nb—Sn phase diagram and the absence of any really clear and logical explanation for the instability of the superconducting properties of Nb_3Sn, that we undertook the present investigation.

TABLE 2. Tin Content of the Alloys Studied

Charge		Analysis		Charge		Analysis	
at.%	wt.%	at.%	wt.%	at.%	wt.%	at.%	wt.%
5	6.30	—	—	28	33.2	23.3	28.0
10	12.43	—	—	30	35.38	26.3	31.3
15	18.40	9.3	11.6	32	37.59	31.4	36.9
20	24.21	16.9	20.6	33.33	38.98	26.9	32.0
25	29.87	22.1	26.6	37.5	43.39	34.9	40.7
26	31.0	22.2	26.8	40	46.0	37.5	43.4

Experiments

The original materials for preparing the alloys were niobium rods (99.6% pure) and tin bars (99.9995% pure). Alloys containing up to 40 at.% Sn were melted in an arc furnace with a tungsten electron and a copper hearth (water-cooled) in an argon medium, the latter being purified by means of a Ti-Zr getter. The alloys were remelted up to six times with reversal of the bars so as to ensure uniformity of composition. During the melting some of the tin evaporated, so that the final composition of the alloy differed from the original. The composition (Table 2) was checked by weighing the resultant material and by chemical analysis.

The cast alloys were annealed at 700, 800, 1000, 1400 and 1800°C. For annealing at 700, 800, and 1000°C the alloys were sealed into quartz ampoules after evacuating these and refilling with argon; the annealing time at these temperatures was 100 h. After annealing, the samples were water-quenched, breaking the ampoules. In addition to this, a sample of stoichiometric Nb_3Sn was annealed in a quartz ampoule at 750°C for 350 h.

The annealing at 1400 and 1800°C was carried out in a TVV-4 furnace in an argon atmosphere for 35 and 5 h respectively. Then the samples were transferred to a quenching apparatus, heated again to the same temperature in an argon atmosphere, held for 15 min, and quenched in castor oil.

The main method of investigation was that of x-ray structural analysis. Microstructural analysis supplemented this. The x-ray diffraction pictures were taken with rotating powder samples in cylindrical cameras 142.8 and 57.3 mm in diameter, using unfiltered chromium radiation. The microsections for microanalysis were prepared mechanically. As etchant we used an $HF-HNO_3-HCl-H_2O$ mixture in the ratio 1:2:5:20.

Results and Discussion

The x-ray structural analysis of alloys containing from 16.9 to 37.5 at.% Sn after annealing at 700 to 1800° showed that these alloys contained either exclusively, or at any rate predominantly, a phase with a crystal structure of the β-W (A15) type, i.e., Nb_3Sn. Thus the conclusion of [18-20, 23] to the effect that Nb_3Sn was only stable above 775 to 860°C receives no support.

The x-ray structural and microstructural investigations showed that Nb_3Sn had an appreciable concentration range of homogeneity; the x-ray diffraction pictures preliminarily taken in the small camera (57.3 mm) indicated that the lattice constant of Nb_3Sn varied little with composition. Precision measurements of the lattice constant of Nb_3Sn were made on the basis of x-ray diffraction pictures taken in the large camera (142.8 mm diameter), using the (421) reflection with a glancing angle of the order of 83°. These showed that the lattice constant varied linearly with composition and increased with increasing tin content (Table 3). The results of lattice-constant measurements carried out on the β-Nb_3Sn phase in alloys quenched after annealing at various temperatures yielded the solubility of tin and niobium in Nb_3Sn at 700, 1000,

TABLE 3. Lattice Parameter (a, Å) of Nb₃Sn as a
Function of Composition and Annealing Temperature

% tin (by analysis)		700°	1000°	1400°	1800°
at.%	wt.%				
9.3	11.6	5.2789	5.2815	5.2800	5.2805
16.9	20.6	5.2810	—	—	—
22.1	26.6	5.2837	—	—	—
22.2	26.8	5.2832	—	—	—
23.3	28.0	5.2842	—	—	—
26.3	31.3	5.2858	—	—	—
31.4	36.9	5.2816	—	—	—
26.9	32.0	5.2864	—	—	5.2858
34.9	40.7	5.2869	5.2878	5.2884	5.2842
37.5	43.4	5.2864	5.2873	—	

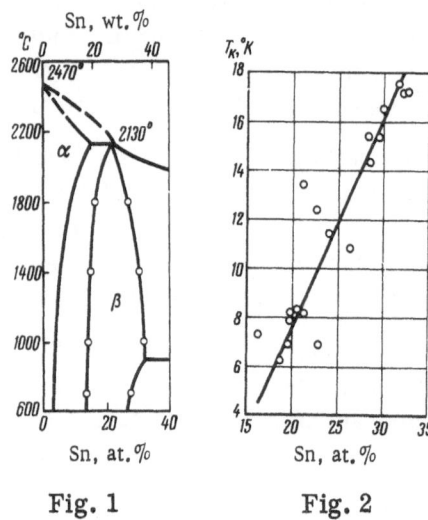

Fig. 1   Fig. 2

Fig. 1. Niobium side of the phase diagram of the Nb–Sn system.

Fig. 2. Superconducting transformation temperature of Nb₃Sn as a function of composition

1400, and 1800°C and thus facilitated the plotting of the range of homogeneity of the β phase
(Fig. 1). It was found that the curve representing the solubility of niobium in Nb₃Sn was monotonic, while that representing the solubility of tin in Nb₃Sn had a bend at 900°C. The maximum
width of the β region was 18 to 19% at 900°C. In the diagram of Fig. 1 the horizontal corresponding to the peritectic equilibrium $\alpha + L \rightleftarrows \beta$ is drawn at 2130°C from the results of [22, 32],
the liquidus line is taken from [32], and the line presenting the solubility of tin in niobium is
taken from the results of [1, 20, 22]. The following represent our own data regarding the solubility of tin in Nb₃Sn:

Temperature, °C	700	1000	1400	1800
Solubility, at.%				
maximum	27.2	31.0	30.0	26.3
minimum	13.0	14.2	15.1	16.0

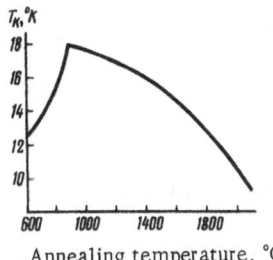

Fig. 3. Theoretical dependence of the critical temperature of Nb$_3$Sn on annealing temperature.

The nonmonotonic nature of the temperature dependence of the solubility of tin in Nb$_3$Sn is evidently due to the fact that at 900°C the solubility curve intersects the peritectic horizontal. The point of intersection (at 32% Sn and 900°C) is a nonvariant point, and the solubility curve accordingly experiences a sharp break.

We considered that the characteristic shape of the solubility curve was the reason for the variation in the superconducting properties of the Nb$_3$Sn phase on varying the annealing temperature. It is known from published data [11, 29, 32, 33, 36] that an increase of the tin concentration in Nb$_3$Sn leads to a rise in the critical current density and critical temperature. However, the results of [11, 29, 32, 33] cannot be used to plot a clear relation between the critical temperature and composition of the compound Nb$_3$Sn. There is a certain spread in the values, apparently associated with the inhomogeneity of the samples and the inaccurate determination of composition. The relation between the critical temperature and composition was not determined in [40]; however, the authors gave a critical temperature and lattice constant for every sample (Table 4). Using our own relation between lattice constant and composition, we determined the composition of each of the samples of [40] and plotted a graph relating the critical temperature to the composition (Fig. 2). The relationship was very nearly linear.

It is clear that the critical temperature for the diffusion layer of Nb$_3$Sn obtained at each specified annealing temperature is determined by the limiting concentration of tin in the β–Nb$_3$Sn phase at this particular temperature. By using the T_K/Sn concentration graph (Fig. 2) and the solubility curve of tin in Nb$_3$Sn (Fig. 1), we may therefore plot a theoretical curve relating T_K to the annealing temperature of Nb$_3$Sn wire. This curve is shown in Fig. 3. On raising the annealing temperature to 900°C the critical temperature increases sharply to 18°K; for annealing temperatures above 900° it gradually falls. This is exactly the kind of variation in critical temperature which is observed experimentally.

Conclusions

1. The compound Nb$_3$Sn is apparently stable down to room temperature and has a wide range of homogeneity. The width of this range increases on raising the temperature to 900°C and at this temperature reaches 18 to 19 at.%; on further raising the temperature it contracts.

2. The critical superconducting temperature of Nb$_3$Sn rises linearly with increasing concentration of tin in the Nb$_3$Sn.

TABLE 4. Lattice Constant and Critical Temperature of Various Nb$_3$Sn Samples Given in [40]

a, Å	T_K, °K	a, Å	T_K, °K	a, Å	T_K, °K
5.289$_2$	17.1	5.287$_5$	15.4	5.282$_5$	7.9
5.289$_3$	17.1	5.285$_9$	10.8	5.283$_2$	8.2
5.288$_6$	17.5	5.282$_5$	8.2	5.282$_8$	8.3
5.287$_7$	16.5	5.283$_9$	12.4	5.282$_3$	6.9
5.284$_1$	6.9	5.284$_8$	11.4	5.283$_2$	13.4
5.287$_0$	14.3	5.280$_6$	7.3	5.281$_9$	6.2
5.286$_8$	15.4				

3. The change in the critical temperature of Nb_3Sn wire following a change in annealing temperature is due to the nonmonotonic temperature dependence of the solubility curve of tin in Nb_3Sn; the same applies to the other superconducting parameters.

LITERATURE CITED

1. M. I. Agafonova, V. V. Baron, and E. M. Savitskii, Izv. Akad. Nauk SSSR, Otd. Tekh. Nauk, Metallurgiya i Toplivo, No. 5, p. 139 (1959).
2. B. T. Matthias, T. H. Geballe, S. Geller, and E. Corenzwit, Phys. Rev., 95(6):1435 (1954).
3. S. Geller, B. T. Matthias, and R. Goldstein, J. Am. Chem. Soc., 77(6):1502 (1955).
4. R. M. Bozorth, A. J. Williams, and D. D. Davis, Phys. Rev. Lett., 5(4):148 (1960).
5. J. E. Kunzler, E. Buehler, F. S. L. Hsu, and J. H. Wernick, Phys. Rev. Lett., 6(3):89 (1961).
6. V. D. Arp, R. H. Kropschot, J. H. Wilson, W. F. Love, and R. Phelan, Phys. Rev. Lett., 6(9):452 (1961).
7. J. O. Betterton, R. W. Boom, G. D. Kneip, R. E. Worsham, and C. E. Roos, Phys. Rev. Lett., 6(10):532 (1961).
8. D. Cline, R. H. Kropschot, V. D. Arp, and J. H. Wilson, High Magnetic Fields, Technol. Press, Cambridge, Mass. (1962), p. 580.
9. H. R. Hart, J. S. Jacobs, C. L. Kolbe, and P. E. Lawrence, ibid., p. 584.
10. E. J. Saur and J. P. Wurm, ibid., p. 589.
11. J. J. Hanak, G. D. Cody, P. R. Aron, and H. C. Hitchcock, ibid., p. 592.
12. F. Lange, Monatsberichte der Deutschen Akademie der Wissenschaften, 1:408 (1959).
13. E. Saur and P. Schult, Z. Physik, 167:170 (1962).
14. V. S. Kogan, A. I. Krivko, B. G. Lazarev, L. S. Lazareva, A. A. Matsakova, and O. N. Ovcharenko, Fiz. Met. Metallov., 15(1):143 (1963).
15. C. L. Kolbe and C. H. Rosner, Metallurgy of Advanced Electronic Materials, Vol. 19, Interscience Publishers, New York (1963), p. 17.
16. E. Buehler, J. H. Wernick, K. M. Olsen, F. S. L. Hsu, and J. E. Kunzler, ibid., p. 105.
17. L. L. Wyman, J. R. Cuthill, G. A. Moore, J. J. Park, and H. Yakowitz, J. Res. National Bureau of Standards, Physics and Chemistry, 66A(4):351 (1962).
18. T. B. Reed, H. C. Gatos, W. J. La Fleur, and J. T. Roddy, Superconductors, Interscience, New York (1963), p. 143.
19. R. Enstrom, T. Courtney, G. Pearsall, and J. Wulff, Metallurgy of Advanced Electronic Materials, Vol. 19, Interscience, New York (1963), p. 121.
20. R. E. Enstrom, G. W. Pearsall, and J. Wulff, J. Metals, 16(1):97 (1964).
21. H. W. Schadler and H. S. Rosenbaum, J. Metals, 16(1):97 (1964).
22. T. G. Ellis and H. A. Wilhelm, J. Less-Comm. Metals, 7(1):67 (1964).
23. H. J. Levinstein and E. Buehler, Trans. Met. Soc. AIME, 230(6):1314 (1964).
24. R. E. Enstrom, G. W. Pearsall, and J. Wulff, Bull. Am. Phys. Soc., Ser. 2, 7(4):323 (1962).
25. T. B. Reed and H. C. Gatos, J. Appl. Phys., 33(8):2657 (1962).
26. D. J. Van Ooijen, J. H. N. Van Vucht, and W. F. Druyvesteyn, Phys. Lett., 3:128 (1962).
27. M. L. Picklesimer, Appl. Phys. Lett., 1(3):64 (1962).
28. R. E. Enstrom, G. W. Pearsall, and J. Wulff, Appl. Phys. Lett., 3(5):81 (1963).
29. T. B. Reed, H. C. Gatos, W. J. La Fleur, and J. H. Roddy, Metallurgy of Advanced Electronic Materials, Vol. 19, Interscience, New York (1963), p. 71.
30. J. H. N. Van Vucht, H. A. C. M. Bruning, and H. C. Donkersloot, Phys. Lett., 7(5):297 (1963).
31. A. H. Gomes de Mesquita, C. Langereis, and J. I. Leenhouts, Philips Res. Reports, 18(5):377 (1963).
32. L. J. Vieland, RCA Review, 25(3):366 (1964).
33. J. J. Hanak, K. Strater, and G. W. Cullen, ibid., p. 342.

34.   R. G. Maier and G. Wilhelm, Z. Naturforsch., 19a(3):399 (1964).

35.   F. J. Bachner, H. C. Gatos, and M. D. Banus, Trans. Met. Soc. AIME, 233(1):227 (1965).

36.   K. Knox, H. Levinstein, and E. Buehler, cited in [21].

37.   J. H. N. Van Vucht, H. A. C. M. Bruning, H. C. Donkersloot, and A. H. Gomes de Mesquita, Philips Res. Reports, 19(5):407 (1964).

38.   H. G. Jansen, Z. Physik, 162(3):275 (1961).

39.   T. Courtney, G. Pearsall, and J. Wulff, J. Metals, 16(1):97 (1964).

40.   T. H. Courtney, G. W. Pearsall, and J. Wulff, Trans. Met. Soc. AIME, 233(1):212 (1965).

41.   J. J. Hauser, Phys. Rev. Lett., 13(15):470 (1964).

42.   B. W. Batterman and C. S. Barrett, Phys. Rev. Lett., 13(13):390 (1964).

43.   J. R. Ogren, T. G. Ellis, and J. F. Smith, Acta Crystallographica, 18(5):968 (1965).

44.   D. B. Montgomery, Bull. Am. Phys. Soc., Ser. 2, 10(3):359 (1965).

DEPOSITION OF Nb_3Sn ON A WIRE FROM THE GAS PHASE

V. I. Arkharov, B. S. Borisov, A. I. Moiseev, and T. A. Ugol'nikova

A method of continuously depositing the superconducting compound Nb_3Sn on a moving substrate is devised. As original materials for obtaining the Nb_3Sn, $NbCl_5$, and $SnCl_4$ are chosen and these are fed separately to the wire. In an apparatus based on this principle, wires with a continuous, single-phase layer of Nb_3Sn firmly attached to the substrate (Pt or Nichrome) are obtained. On changing the deposition conditions, the thickness of the layer changes from one to several microns.

The prospects of obtaining high magnetic fields by means of the superconducting compound Nb_3Sn have evoked a number of papers seeking means of using this compound in solenoids [1-8], since by itself the compound is extremely brittle and wire cannot be made from Nb_3Sn in any ordinary manner.

The first of the methods proposed was that of forming Nb_3Sn by the heat treatment of solenoids made from material containing niobium and tin. Kunzler et al. [2] used a niobium tube filled with a mixture of niobium and tin powders, drawn to the required diameter.

Alekseevskii and Mikhailov [6] used tinned niobium wire as original material.

In other experiments attempts were made to prepare a flexible conductor earlier coated with a thin layer of Nb_3Sn [5, 7-10].

The most promising was the method of obtaining a superconducting wire by depositing a layer of Nb_3Sn from the gas phase on a moving refractory substrate, as proposed by Hanak [8-10]. This method was based on the simultaneous hydrogen reduction of niobium and tin chlorides on a hot substrate and yielded wire or strip with a homogeneous layer of Nb_3Sn firmly attached to the substrate. The thickness of the layer was given by the conditions of deposition; it was constant the whole length of the wire.

In Hanak's apparatus the chloride vapor passed into a reaction chamber through which a refractory wire heated to between 800 and 1200°C by a current moved. At this temperature the Nb_3Sn was deposited on the wire at a fair rate. The choice of temperature for the reaction chamber was a difficult problem. The chamber had to be heated sufficiently to prevent condensation of the chlorides, but at the same time its temperature had to be lower than that of the wire, so as to prevent the formation of Nb_3Sn on the chamber walls.

However, at this temperature (about 700°C), the vapor of the volatile niobium pentachloride was rapidly reduced by hydrogen to involatile niobium trichloride, which blocked up the reaction chamber in the form of a loose black powder.

Consideration of the thermodynamics of the reduction process enabled Hanak to find a safe temperature range by adding hydrogen chloride. The addition of hydrogen chloride raises the temperature for the formation of Nb_3Sn and reduces that corresponding to the stability of niobium trichloride, thus retarding the harmful processes. By introducing controlled amounts of hydrogen chloride, Hanak was able to keep the temperature of the reaction chamber between 720 and 790° without seriously blocking the chamber, for a period sufficient to obtain a conductor 300 m long [9]. Hanak proposed increased the working period of the apparatus to 90 h, giving section of wire up to 1000 m long [10].

The niobium and tin chloride vapors for the Hanak-type apparatus were obtained by a) heating niobium and tin chlorides, b) chlorinating sintered Nb_3Sn [1], and c) separately chlorinating tin and niobium powders [10]. Hanak regarded $NbCl_4$ and $SnCl_2$ as the best combination of chlorides [10].

Our own problem was to create an apparatus for the deposition of Nb_3Sn from the gas phase, capable of operating for a long period with blocking up the reaction chamber.

We first undertook a detailed consideration of the interactions taking place in the reaction chamber. The main source of blocking was the formation of the involatile niobium trichloride. There were two processes leading to the formation of this substance: incomplete reduction of the pentachloride by hydrogen

$$NbCl_5 + H_2 = NbCl_3 + 2HCl \tag{1}$$

and reduction of the niobium pentachloride by stannous chloride

$$NbCl_5 + SnCl_2 = NbCl_3 + SnCl_4. \tag{2}$$

The velocity of the two reactions becomes substantial at a temperature of about 200°C and increases with rising temperature [11, 12].

Hence in order to prevent the formation of niobium trichloride in the reaction chamber the temperature of the latter should be kept below 200°C. However, in order to prevent the condensation of the original chlorides, these must have a reasonably high vapor tension at this temperature. Of the various niobium chlorides this requirement is best satisfied by the pentachloride, and of the tin chlorides by the stannic tetrachloride. We therefore chose Nb_2Cl_5 and $SnCl_4$ as a working pair. This choice not only sharply reduces the rate of formation of niobium trichloride by reaction (1) but altogether prevents its formation by reaction (2).

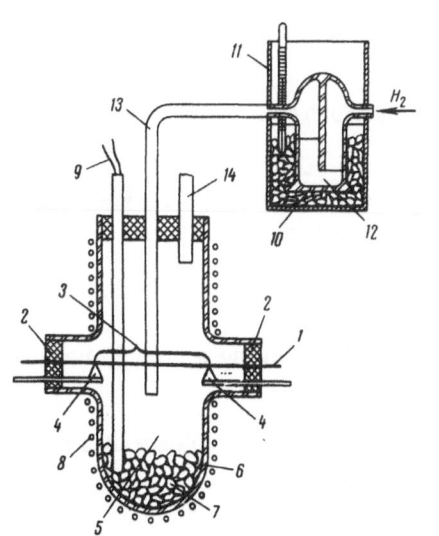

Fig. 1. Apparatus for depositing Nb_3Sn from the gas phase on a wire.

In order to reduce the amount of niobium trichloride further, we decided to minimize the time spent together by hydrogen and niobium pentachloride by feeding these separately to the place of deposition. Hydrogen was taken as carrier gas for the $SnCl_4$ vapor as well, since no interaction took place between these up to 350°C [13].

On the basis of the foregoing analysis and some preliminary experiments, we devised an original method and the corresponding apparatus for depositing a layer of Nb_3Sn on a moving substrate.

A schematic diagram of the apparatus for continuous deposition on a moving wire appears in Fig. 1. The wire 1, passing through the gland 2, moves forward through the reaction space at a specified velocity. The working part of the wire 3 is heated to a temperature of the order of 1000°C by an electric current supplied through the sliding contacts 4. A reservoir 6 contain-

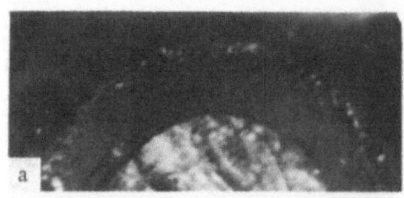
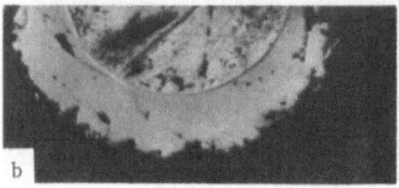

Fig. 2. Layer of Nb_3Sn deposited on a platinum wire 0.3 mm in diameter as seen in polarized and unpolarized light (a and b).

ing solid niobium pentachloride 7 communicates with the reaction chamber 5. The chamber and reservoir are heated by a single furnace 8 to a temperature of 120 to 170°C, regulated by a thermocouple 9. The vapor formed by the heating of the niobium pentachloride passes directly to the surface of the wire under treatment. The hydrogen, passing through the reservoir 10 held in a thermostat at 0°C, is saturated with stannic chloride vapor 12 and passes through the tube 13 to the reaction space and thus to the surface of the working part of the wire, on which a layer of Nb_3Sn is formed. The spent gases pass out through the tube 14.

Under the conditions described, the involatile niobium trichloride is formed in very small quantities, since the dangerous temperature zone occupies a narrow region near the heated wire. No formation of niobium trichloride on the walls of the chamber occurs at the temperature chosen. Thus the service life of our apparatus is not determined by blockage with involatile products.

The apparatus was tested for continuous deposition on a moving platinum or Nichrome wire 0.3 mm in diameter. The rate at which the wire passed through the apparatus was 4 to 10 m/h. The diameter of the receiving drum was 80 mm, and the rate of hydrogen feed 1.5 to 2 liters/h. In this way a homogeneous layer with the β-W structure was obtained on the wire; the thickness of the layer, determined from the weight increment, was 1 to 6 μ, increasing with rising wire temperature (the thickness of the deposited layer also depended on the temperature of the niobium pentachloride).

The temperature of the wire was varied from 800 to 1200°C. The upper limit to the temperature of the wire was set by its heat resistance.

The wire covered with the Nb_3Sn layer had a smooth, even surface. After winding the wire onto a cylinder 6 mm in diameter, continuous transverse cracks developed in the surface layer (for a platinum substrate). In samples with a Nichrome substrate the cracks extended over part of the circumference only; for a very thick Nb_3Sn layer cleavages or fissures appeared on the surface.

The same apparatus could be used to deposit Nb_3Sn on a stationary substrate and obtain a deposit 100 μ thick in 20 min.

A photograph of the cross section of one such sample is presented in Fig. 2. We see that the Nb_3Sn layer constitutes a dense, pore-free deposit, firmly attached to the substrate.

Conclusions

1. We have proposed a method and designed an apparatus for obtaining a homogeneous layer of superconducting Nb_3Sn on a moving substrate.

2. The method ensures prolonged operation of the apparatus without blockage by involatile products.

LITERATURE CITED

1. J. J. Hanak, Metallurgy of Advanced Electronic Materials, Vol. 19, Interscience, New York (1963), p. 161.
2. J. E. Kunzler, E. Buehler, F. S. Hsu, and J. H. Wernick, Phys. Rev. Lett., 6(3):89 (1961).
3. A. C. Barber and P. H. Morton, Nature, 198(4875):82 (1963).
4. C. L. Kolbe and C. L. Rosner, Metallurgy of Advanced Electronic Materials, Vol. 19, Interscience, New York (1963), p. 17.
5. E. J. Saur and J. B. Wurm, High Magnetic Fields, Cambridge, Mass. (1962), p. 589.
6. N. E. Alekseevskii and N. N. Mikhailov, Zh. Éksp. Teor. Fiz., 41 (Issue 6, No. 12):1810 (1963).
7. M. D. Banus, H. C. Gatos, M. C. Lavine, and T. B. Reed, Metallurgy of Advanced Electronic Materials, Vol. 19, Interscience, New York (1963), p. 151.
8. Elektronika, No. 30, p. 42 (1962).
9. J. J. Hanak, G. D. Cody, P. R. Aron, and H. C. Hitchcock, Intern. Conf. on High Magnetic Fields, Cambridge (1961), p. 592.
10. J. J. Hanak, K. Strater, and C. W. Gullen, RCA Review, Vol. 23, No. 342 (1964).
11. I. S. Morozov and Li Chi Fa, Zh. Neorg. Khim., 8(1):2733 (1963).
12. H. Brauer, Handbook of Preparative Inorganic Chemistry, Academic Press, New York (1963).
13. Ya. I. Gerasimov, A. E. Krestovnikov, and A. S. Shakhov, Chemical Thermodynamics in Nonferrous Metallurgy [in Russian], Vol. II, Metallurgizdat (1961).

STRUCTURE AND SUPERCONDUCTING PROPERTIES OF SOME SUPERCONDUCTING ALLOYS OF THE Nb—Ga SYSTEM

L. F. Myzenkova, V. V. Baron, and E. M. Savitskii

The relation between the critical current and the magnetic field is studied for alloys of the Nb—Ga system in the range of solid solutions. The transformation temperature of various Nb—Ga alloys is measured. The system contains no compounds having a T_K exceeding the critical temperature of Nb_3Ga (14.5°K).

In studying the structure and properties of superconducting alloys, the greatest attention has so far been paid to alloys of niobium. In addition to the solid solutions of niobium with transition metals possessing high superconducting characteristics, niobium forms superconducting compounds with tin, aluminum, and gallium [1–3]; the superconducting transformation temperatures of these alloys are amongst the highest yet known. Such compounds (with cubic lattices of the Cr_3Si type) are formed in particular in alloys of niobium with elements belonging to the B subgroups of the periodic system. Compounds with the Cr_3Si structure have much higher superconducting characteristics than other compounds.

In view of this it is particularly interesting to study alloys of niobium with elements of low melting points.

The phase diagram of the Nb—Ga system was plotted earlier [1] on the basis of microstructural and thermal analysis together with x-ray diffraction and measurements of microhardness (Fig. 1). A continuous Nb-base solid solution is formed in this system. The solubility of gallium in niobium at 800° is about 8 wt.%. In addition to the earlier-known compound Nb_3Ga (20.1 wt.% Ga), three more compounds occur in the system. The compound Nb_3Ga, a superconductor with a superconducting transformation temperature of 14.5°K, has a structure of the Cr_3Si type. The following compounds are formed by peritectic reactions: Nb_5Ga_3 (31.08 wt.% Ga, tetragonal structure of the W_5Si_3 type), $NbGa_3$ (69.29 wt.% Ga, tetragonal structure of the $TiAl_3$ type), and (apparently) Nb_2Ga_3 (about 51 wt.% Ga). A pseudoeutectic is formed on the gallium side.

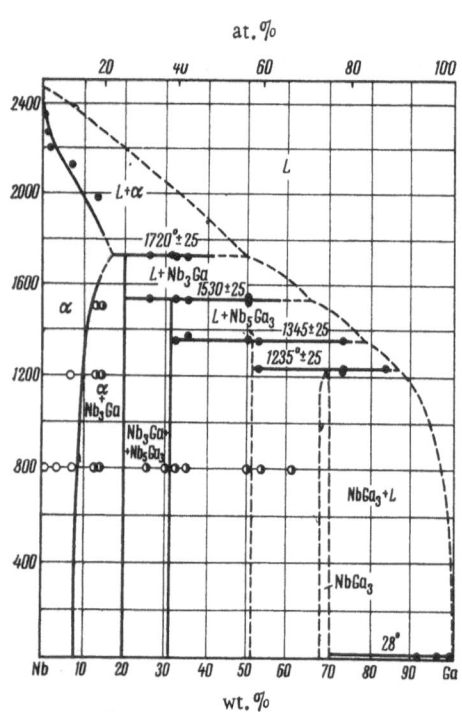

Fig. 1. Phase diagram of the Nb—Ga system.

130

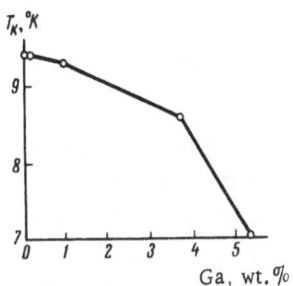

Fig. 2. Superconducting transformation temperature as a function of gallium content for Nb–Ga alloys in the solid-solution range.

We studied the dependence of the critical current on the magnetic field for Nb–Ga alloys in the range of solid solutions and measured the superconducting transformation temperature.

Both T_K and I_K were measured on wire samples. The alloys were melted in an arc furnace with an expendable electrode on a water-cooled copper hearth in an atmosphere of purified helium. Wire 0.25 mm in diameter was prepared from alloys in the region of Nb-base solid solution by cold rolling and drawing.

Figure 2 shows the dependence of the superconducting transformation temperature of Nb–Ga alloys on the gallium content over the range of solid solutions. The temperature of the superconducting transformation was determined from the change in the magnetic permeability of the sample. We see that the superconducting temperature of the alloy falls with increasing gallium content; for alloys with 5.4 wt.% Ga it equals 7°K.

In addition to the range of solid solutions, we studied alloys of the Nb–Ga system over the range of concentrations up to 96.7% Ga. Data relating to the transformation temperature of these alloys are presented in Table 1. The two-phase alloy, comprising a solid solution and the compound Nb_3Ga, passes into the superconducting state at 9°K. For the compound Nb_3Ga the critical temperature is 14.2°K. The next two-phase alloy (compounds Nb_3GA and Nb_5Ga_3) gives two steps on the transformation curve, at 13.5 and 7.5°. The remaining alloys have a constant transformation temperature, right up to 96.7% Ga.

Our results on the relation between the transformation temperature and the gallium content in the range of solid solutions are analogous to those relating to the effect of tin content on the T_K of niobium in the Nb–Sn system published in [4, 5]. On adding even 0.66 at.% Sn to niobium the critical temperature fell to 8.8°K; 8 at.% of Sn reduced it to 5.6°K.

A fall in the T_K of vanadium in the range of V–Ga solid solutions has also been reported in the literature [6]. The results of our own investigations and the published data presented confirms Matthias' conclusion that when solid solutions are formed between nontransition and transition metals the critical temperature will fall, independently of the concentration of va-

TABLE 1. Superconducting Transformation Temperatures of Niobium–Gallium Alloys

Composition, wt.% Ga	T_K, °K	Phase composition
Nb	9.5	Nb
0.06	9.4	Solid solution of Ga in Nb
1.0	9.3	"
3.76	8.6	"
5.4	7.0	"
13.26	9.05	Solid solution of Ga in Nb + Nb_3Ga
20.1	14.2	Nb_3Ga
30.8	13.5 and 7.5	$Nb_3Ga + Nb_5Ga_3$
35.6	7.2	Nb_5Ga_3
38.7	7.4	Nb_5Ga_3
42.59	7.5	Nb_5Ga_3
47.56	7.4	Nb_5Ga_3
50.0	7.6	—
61	7.3	—
69.29	7.6	$NbGa_3$
96.7	7.3	$NbGa_3$ + Ga

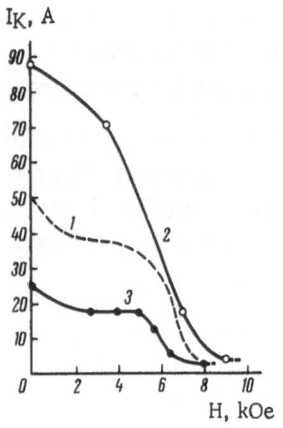

Fig. 3. Critical current as a function of magnetic field for Nb—Ga alloys. Gallium content, wt.%: 1) 0; 2) 1; 3) 5.4.

lence electrons per atom, in contradistinction to solid solutions of transition metals, in which T_K is an oscillatory function of the number of valence electrons per atom.

In our own investigations into the properties of the Nb—Ga alloys, the measurements of critical current density for various magnetic fields were carried out on short samples, wire 0.25 mm in diameter and 100 mm long. The samples were placed in a transverse magnetic field. All the measurements were made at 4.2°K. The maximum magnetic field of the apparatus was 26 kOe. The transformation from the superconducting to the normal state was determined by reference to the appearance of a 20-mV voltage drop in the sample.

The field dependence of the critical current for alloys with various gallium contents is shown in Fig. 3. We see that very small quantities of gallium raise the critical current of niobium in the low-field region. There is also a certain rise in the critical magnetic field for an alloy containing 1% Ga. Raising the gallium content to 5.4 wt.% reduces the critical current density and the critical magnetic field.

Thus our experiments show that the superconducting transformation temperature falls for solid solutions of gallium in niobium.

In studying the transformation temperature of Nb—Ga alloys no compounds with a critical temperature higher than that of Nb_3Ga (14.5°K) were found.

LITERATURE CITED

1. V. V. Baron, L. F. Myzenkova, E. M. Savitskii, and E. I. Gladyshevskii, Zh. Neorg. Khim., 9(9):1 (1964).
2. J. K. Hulm and R. D. Blaugher, Phys. Rev., 123(5):1569 (1961).
3. B. T. Matthias, Rev. Mod. Phys., 35:1 (1963).
4. J. H. N. Van Vucht, D. J. Ooijen, and H. A. C. M. Bruning, Philips Res. Reports, 20(2):136 (1965).
5. G. E. Telentyuk et al., in: "Metallography and Metallophysics of Superconductors" [in Russian], Izd. "Nauka" (1965), p. 83.
6. J. H. N. Van Vucht, H. A. C. M. Bruning, H. C. Donkersloot, and A. H. Gomes, Philips Res. Reports, 19(5):407 (1964).

SUPERCONDUCTING COMPOUNDS OF VANADIUM
Yu. V. Efimov

The crystal structure of vanadium compounds and the corresponding superconducting transformation temperature are considered as functions of the position of the second component in the periodic system. Among binary vanadium compounds with unknown T_K only V_3Al and V_3In can have a high critical temperature. The search for new superconducting compounds is most promising in ternary systems in which compounds of the Cr_3Si are formed. The alloying of binary vanadium compounds with this kind of structure and a high T_K always reduces the latter. When continuous solid solutions are formed between superconducting compounds of the Cr_3Si type the critical temperature changes smoothly with the composition of the alloys and may be expressed by empirical exponential functions. In such systems (with certain ranges of concentration) there is an approximately linear relationship between T_K and the lattice constant of the solid solutions.

One of the fundamental problems in the metallography of superconductors is that of establishing criteria for the appearance of superconductivity in alloys and compounds, and studying their superconducting properties in conjunction with the principal physicochemical constants of the components [1]. However, it is as yet impossible to make a theoretical prediction as to the appearance of superconductivity in new alloys and compounds. There are only a few empirical conditions essential to the existence of superconductivity. The electron concentration of almost all known superconductors is greater than 2 but less than 8 electrons per atom [2, 3]. Matthias et al. [2, 3] established a certain correlation between the critical temperature of superconductors and their electron concentration. Electron concentrations favorable to high critical temperatures are rather different for compounds with different crystal structures. The critical temperature of the elements also varies with their atomic volumes and atomic masses. The effect of crystal structure on superconductivity appears more sharply in compounds of transition metals. Particularly favorable for superconductivity are compounds with cubic structures of the Cr_3Si, NaCl, and $MgCu_2$ types and also tetragonal σ phases [1, 2]. The highest critical temperature occurs for compounds with the Cr_3Si structure [1, 2].

The present problem was to consider the crystal structure and superconducting transformation temperature of binary vanadium compounds as functions of the position of the second component in the Mendeleev periodic table, and also to analyze the variations in the transformation temperature of certain three-component compounds of vanadium.

The formation of compounds is ultimately conditioned by the disposition of the interacting elements in the periodic system. The type of compound, its crystal structure, stability, and properties depend on the differences in the structure of the outer electron shells, the atomic radii, the electronegativity, and other physicochemical properties of the components determining the type and strength of the interatomic bonds in the crystals [5-7]. However, it is still only possible to predict the formation of specific types of compounds in a few individual cases with the help of atomic radii, electronegativities, and thermodynamic characteristics [2, 7-9].

Compounds with a purely metallic bond are formed in systems of typical metals with comparatively similar properties. For an electronegativity difference (ΔE_n) of under 2 units in systems of metals and nonmetals, covalent compounds are formed [5, 10]. A great difference in electronegativity ($\Delta E_n > 2$) leads to the formation of simple chemical compounds with an ionic bond [10]. The probability of the formation of compounds increases with increasing difference in the electronegativity of the elements [11].

On considering the order of electronegativity in the elements of the periodic system [12] we see that in the majority of systems compounds exist linking vanadium with more electronegative elements, starting from Re [13, 14]. Among these elements only a few electronegative metals form no compounds with vanadium; these fail to mix with vanadium in the liquid state over a wide range of concentrations. Of elements more electropositive than vanadium, only Zr, Hf, and Ta form compounds with the latter. In binary systems of vanadium with electropositive metals and with Group III transition metals (Sc, Y, r.e.m.) there are no compounds. Nor does vanadium form compounds with neighbors in the electronegativity series: Nb, W, Cr, Mo, and the actinides. The number of compounds in binary systems of vanadium in general increases with increasing number of the group characterizing the second component, and with increasing difference between the structures of the outer electron shells of vanadium and the second component of the alloy. There is no very clear tendency of the number of compounds to diminish (particularly in systems involving elements of Groups VIII and VIB) with increasing atomic number of the second component within each group of the periodic system.

The types of crystal structure and the possible ranges of homogeneity of the compounds in systems incorporating vanadium and transition metals are shown in Fig. 1.

Laves phases (Fig. 1b) are only formed in systems of vanadium with more electropositive transition metals (Zr, Hf, Ta) [13, 14, 15]. The positive size factor in these systems [11] lies in the range of values characterizing Laves phases (10 to 50%) [8, 16]. The compounds ZrV_2, HfV_2, TaV_2 have a cubic structure of the $MgCu_2$ type [15]. With increasing difference in the structure of the outer electron shells of vanadium and the second components in these phases, the temperature of formation of the phases becomes greater. The existence of closer atomic radii in the V–Ta system evidently promotes the formation of a range of homogeneity in the TaV_2 compound. The probability of the existence of other Laves phases in systems of vanadium with transition metals is very low.

Sigma phases (Fig. 1g) are formed in systems of vanadium with Mn, Fe, Co, Ni and Re, i.e., transition metals having atomic radii similar to that of vanadium ($0.93 \le r_{Me}/r_V \le 1.15$) [15, 17]. In systems with transition metals of the same period we find that, as the atomic number (and hence the difference in the outer electron shells) increases the range of homogeneity of the σ phases moves in the direction of vanadium and there is a rise in the temperature corresponding to the formation of the σ phases. The comparatively large difference in the atomic radii of vanadium and nickel ($r_{Ni}/r_V = 0.926$), somewhat exceeding the limiting value, constitutes the reason for the lower stability of the σ phase NiV (reduction in melting point, polymorphic transformation). The characteristics of the σ phase in the V–Re system (narrow range of homogeneity, only stable at high temperatures) are evidently associated with the great difference in the structure of the outer electron shells of rhenium and vanadium. The difference in the electron

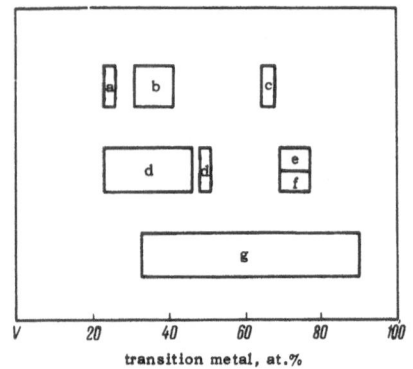

Fig. 1. Types of crystal structure and possible ranges of homogeneity of compounds in systems of vanadium with transition metals.

structure of the interacting elements also has an effect on the conditions of formation of the
σ phases. In systems of vanadium with elements further removed from the latter in the periodic system, σ phases are formed by peritectic reactions, and in systems with similar elements
by the ordering of the solid solutions.

Analysis of the positions of the elements forming σ phases in the periodic system gives
grounds for supposing that, among other binary vanadium systems, the formation of a σ phase
may also occur in the V–Tc system. The sigma phase in this system should be similar to the
σ phase in the V–Re system in its characteristics (range of homogeneity, conditions of formation).

In systems with elements of the VIIA and VIIIA subgroups, ordered phases of the CsCl
type are formed at equiatomic compositions (Fig. 1d). The main condition for the formation
of stable VTc, VRu, and possible VOs phases is the small difference in the atomic radii. For
a size factor exceeding 1% a structure of the CsCl type becomes unstable. For example, in
the V–Fe system a metastable phase of the CsCl type precedes the formation of the σ phase [18].
In addition to this, in systems with elements of the VIIIA subgroup secondary ordered phases
of the CsCl type may be formed in the vanadium-rich region. On passing to elements of the
VIIIB and VIIIC subgroups, phases with a different crystal structure lie in the regions of the
CsCl-type phases.

Vanadium shows a special tendency to form compounds of the Cr_3Si type [15]. Binary
vanadium systems show the greatest number of compounds with this type of structure (Fig. 1a).
In these vanadium takes the place of the A component. In systems with transition metals vanadium forms compounds of the Cr_3Si type with elements of the VIIIB and VIIIC subgroups. It is
characteristic that phases of the Cr_3Si type in systems of vanadium with transition metals have
very narrow ranges of homogeneity and are formed by peritectoid reactions at comparatively
low temperature (700 to 900°).

Over wide ranges of the solid solutions of vanadium in transition metals of the VIIIB and
VIIIC subgroups, ordered phases homogeneous at about 23 to 30 at.% V are formed. In systems
with elements of the VIIIB subgroup these phases have the structure of Cu_3Au (Fig. 1e). In addition to this, in systems with elements of the VIIIC subgroup, at 32 to 35 at.% V yet another
ordered phase with an orthorhombic or tetragonal structure (Fig. 1c) is formed. By analogy
we may expect the formation of a similar phase in the V–Pt system.

As regards its tendency to form compounds with vanadium, the electronegative metal Au behaves in the same way as noble metals of the VIIIC subgroup. This is evidently a consequence of the lower stability of the $(5d)^{10}$ subshell of gold as compared with the electron subshells of Ag and Cu [19]. The difference in the electrochemical nature of Au and the elements of the VIIIC subgroup only affects the formation of ordered phases with a different crystal structure in the region of Au-base solid solutions.

In systems of vanadium with electronegative metals and nonmetals one often encounters different compounds with metallic, covalent, or ionic bonds, and also compounds with a mixed or intermediate type of bond. These vary considerably in their crystal structure.

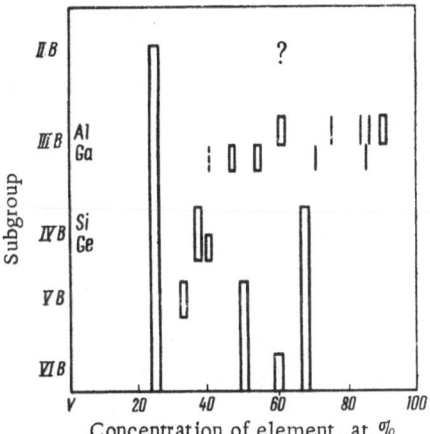

Fig. 2. Typical compositions of
the phases in systems of vanadium
with elements of the IIB to VIB subgroups in the Mendeelev periodic
system.

Typical compositions of the phases in systems of vanadium with elements of the IIB to VIB subgroups are shown in Fig. 2. Owing to the limitations of the present investigations we omitted consideration of compounds of vanadium with B, O, C, N, and H. Borides, carbides, oxides, nitrides, and hydrides are special groups of chemical compounds with sharply individual characters. Many investigations have been directed at these compounds.

Compounds of vanadium with beryllium, VBe_2 ($MgZn_2$ type) and VBe_{12} ($ThMn_{12}$ type) are not typical of other systems of vanadium with electronegative metals. This is evidently because of the special position of Be in the periodic system as compared with other electronegative metals. In the compound VBe_2, in contrast to the Laves phases with transition metals, vanadium, as the element with the greater atomic radius, occupies the position of the A element. In other binary systems of vanadium with electronegative metals no Laves phases are encountered.

In systems of vanadium with Al and Ga a large number of compounds are formed. As regards number of compounds these systems are only inferior to the V–O system. The crystal structure and composition of the compounds of vanadium with Ga or Al, apart from phases of the Cr_3Si type, are not typical of compounds with other electronegative metals and nonmetals.

For all systems of vanadium with elements of the IIB to VIB subgroups of the third to sixth periods, phases of the A_3B composition, in which V plays the part of the A component, are typically formed. A probable exception is the V–Hg system, in which on the basis of the Hildebrand factor [11] we should expect complete immiscibility in the liquid state. These compounds in the systems of vanadium with elements of the IIB to IVB subgroups (third to sixth periods) and also those with As, Sb, and possibly Bi have a crystal structure of the Cr_3Si type. In systems with phosphorus and nonmetals of Group VI, phases of other structures are formed; in these the metallic bond gives way more and more to the covalent and ionic. Compounds of the Cr_3Si type in systems of vanadium with Al and Ga have a characteristic feature: They are formed from the solid solution and have wide ranges of homogeneity. A similar feature clearly characterizes all A_3B phases with elements of the IIB and IIIB subgroups. In systems of vanadium with elements of the the IVB and VB subgroups compounds of the Cr_3Si type are formed by peritectic reactions.

The crystal structure of many compounds of vanadium with elements of the IVB to VIB subgroups is unknown. In these systems, for compounds of the same composition we find a change in crystal structure and type of bond on passing from elements of one group to elements of another, and also on passing from one element to another within the same group (for example, AB_2 phases). The only large group with a common structure among these various compounds is that of the Ni–As phases in systems with elements of the VB and VIB subgroups.

The superconducting transformation temperatures have been studied for a limited number of vanadium compounds (Table 1). Among the binary compounds of vanadium the greatest number of superconducting phases and the highest critical transformation temperatures occur in phases with the Cr_3Si structure. Quite high transformation temperatures characterize the compounds ZrV_2 (Laves phase) and VN, which have the cubic NaCl structure. Among the ternary vanadium compounds only one superconductor is known with a fairly high critical temperature (7.5°K); this is the Zr_3V_3O phase, which is also cubic (Fe_3W_3C type) [2]. Among the binary and ternary compounds of vanadium with other crystal structures there are either no superconductors at all, or else these have very low critical temperatures (under 0.5°K). It is true that ternary compounds have not been very widely studied.

Among the four binary Laves phases of vanadium the greatest critical temperature is that of ZrV_2. The electron concentration is 4.67 electrons/atom. The electron concentrations of TaV_2 (5 electrons/atom) and particularly VBe_2 (3 electrons/atom) differ greatly from the elec-

TABLE 1. Superconducting Transformation Temperature of
Vanadium Compounds

Compound	Type of structure	T_K, °K	Test temperature, °K	Compound	Type of structure	T_K, K°	Test temperature, °K
V_3Si	Cr_3Si	17.1	—	VO	NaCl	—	1.20
V_3Ga	Cr_3Si	16.5	—	V_2C	Hex.	—	1.20
V_3Ge	Cr_3Si	6.01	—	V_5N_2	»	—	1.20
V_3Sn	Cr_3Si	6.00	—	V_2O_3	Al_2O_3	—	1.28
V_3Pt	Cr_3Si	2.83	—	ZrV_2	$MgCu_2$	8.8	—
$(V_{2.67}Ir_{0.33})$ Ir	Cr_3Si	1.39	—	$VOs_{0.40}$	CsCl	—	0.37
V_3Ir	Cr_3Si	—	0.35	VZn_3	Cu_3Au	—	1.02
V_3Sb	Cr_3Si	0.8)	—	$V_{24}Re_{76}$	σ- phase	4.52	—
V_3Au	—	0.74	—	VP	NiAs	—	1.02
V_3Rh	Cr_3Si	0.38	—	VB	CrB	—	1.20
V_3As	Cr_3Si	—	1.02	VB_2	AlB_2	—	1.20
V_3Co	Cr_3Si	—	0.35	V_5Ge_3	Mn_5Si_3	—	1.02
V_2Pt	—	—	1.02	V_5Si_3	—	—	0.35
V_3P	Cr_3P	—	1.00	V_5Si_3	W_5Si_3	—	0.30
VC	NaCl	—	1.20	VGe	$CrGe_2 + x$	—	1.20
VN	NaCl	8.2	—	VSi_2	$CrSi_2$	—	1.20
$V_3Fe_5Si_2$	α-Mn	—	0.37	$ZrV_{0.5}Ni_5$	$MgCu_2$	0.43	—
$V_{2.6}Co_{5.2}Si_2$	α-Mn	—	1.02	Zr_3V_3O	Fe_3W_3C	7.5	—
$V_3Ni_5Ge_2$	α-Mn	—	0.35	$V_{86}O_2N_{12}$	Cub.	5 80	—
Zr_2VCo_3	$MgZn_2$	—	0.35	$V_{79}O_2N_{19}$	»	6.70	—
$NbVCo$	$MgZn_2$	—	1.02	$V_{78}O_8N_{14}$	»	8.20	—

tron concentration favoring a high critical temperature (4.7 electrons/atom) [2, 3]. The replacement of Zr by Hf (element with a large atomic mass) should also apparently lead to a fall in critical temperature [2, 20].

The sigma phases have a transformation temperature of up to 9°K [1, 2, 21]. Among the binary σ phases of vanadium only one superconductor is known, the compound VRe_3 with a critical temperature of 4.52°K [21]. We should expect superconductivity in the σ phase of the V–Mn system.

In compounds of the Cr_3Si type the electron concentrations favorable toward high critical temperatures are equal to roughly 4.5 and 6.5 electrons/atom [3]. Among vanadium compounds with this structure the highest critical temperatures belong to V_3Si (17.1°K) and V_3Ga (16.5°K), which have an electron concentration of 4.75 and 4.5 electrons/atom respectively. In superconducting vanadium compounds of the Cr_3Si type with the same electron concentrations, the replacement of the B element by an element of the same group with a greater atomic mass leads to a fall in critical temperature ($V_3Si \rightarrow V_3Ge \rightarrow V_3Sn$). On this basis we should except that the critical temperature of V_3In would not exceed that of V_3Ga. The phase V_3Al may have a higher critical temperature. The critical temperature of V_3Pb and V_3Tl is evidently low.

The variation in the critical temperature of binary vanadium compounds of the Cr_3Si type with their electron concentration is shown in Fig. 3. A reduction in the electron concentration to 4 electrons/atom (V_3Au) leads to a sharp fall in critical temperature. The electron concentration of V_3Cd (4.25 electrons/atom) is below the favorable value; hence in this compound also we should not expect a high critical temperature. However, the transformation temperature should certainly be a little higher than that of V_3Au.

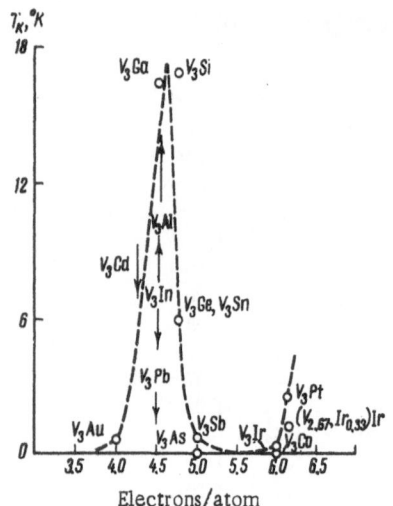

Fig. 3. Superconducting transformation temperature of binary vanadium compounds of the Cr_3Si type.

For an electron concentration of 5 to 6 electrons/atom, compounds of vanadium with other transition metals of the Cr_3Si type have low critical temperatures [2].

Earlier the range of electron concentrations between 5 and 5.25 electrons/atom was generally regarded as one in which no superconductivity existed [3]. Later superconducting compounds of the Cr_3Si type with a critical temperature below 1°K were found in this range [2]. On increasing the electron concentration above 6 electrons/atom, the critical temperature of Cr_3Si compounds rises and for 6.25 to 6.75 electrons/atom reaches quite high values [2]. However, the maximum transformation temperature in this region is about half that of compounds with 4.5 to 4.75 electrons/atom. The electron concentration of all binary vanadium compounds of this type of crystal structure is under 6.33 electrons/atom. The critical temperatures of vanadium compounds with 6.25 to 6.33 electrons/atom is no greater than 3°K. We should hardly expect a high superconducting transformation temperature in V_3Ni, which contains a ferromagnetic metal.

Thus among the binary vanadium compounds of the Cr_3Si type a high transformation temperature is only to be expected from V_3Al and V_3In.

The only prerequisites in seeking new ternary superconducting compounds at the present time are favorable electron concentrations, slightly differing for different types of crystal structure, and a favorable crystal structure (types Cr_3Si, NaCl, $MgCu_2$, Fe_3W_3C; σ phases).

The most interesting are clearly the new ternary compounds of the Cr_3Si type, since alloying binary compounds of this type having high critical temperatures always leads to a fall in the latter [22, 23]. The fall in the critical temperatures of Cr_3Si-type compounds on alloying may be explained by the fact that only for a stoichiometric composition of the compound are the optimum conditions for the development of superconductivity realized. Hence any deviations from stoichiometric composition will lead to a reduction in the critical temperature. The only exception is the Nb_3Sn–Nb_3In system, where small quantities of indium slightly raise the critical temperature of Nb_3Sn [24]. However, the rise in critical temperature is no more than about 0.5°K and may be associated with the diffusion method of obtaining the original Nb_3Sn.

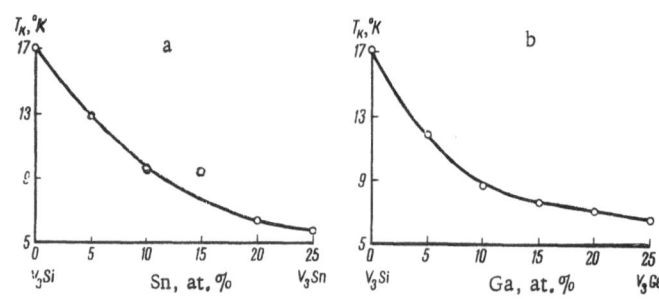

Fig. 4. Critical temperature of the V_3Si–V_3Sn (a) and V_3Si–V_3Ge (b) systems as a function of the composition of alloys.

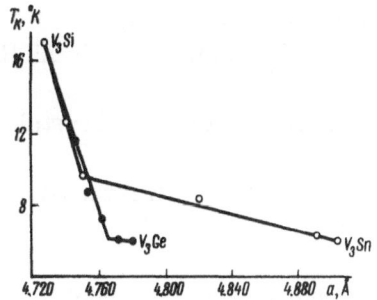

Fig. 5. Critical temperature of V$_3$(Si, Ge) and V$_3$(Si, Sn) solid solutions as a function of the lattice constant.

When a continuous series of solid solutions is formed between superconducting binary compounds of the Cr$_3$Si type, the critical temperature varies smoothly with changing composition. Examples include alloys of V$_3$Si with V$_3$Sn and V$_3$Ge (Fig. 4) [25, 26]. The dependence of the critical temperature on the composition of the alloys in such systems is expressed by the following empirical formula:

$$T_{\text{к}} = T_{\text{к (max)}} \; e^{-A \cdot x} + Be^{C \cdot x},$$

where x is the proportion of the third component, the compound of which with the transition metal has the lower critical temperature in the pseudobinary system under consideration, $T_{K \, (max)}$ is the critical temperature, and A, B, and C are empirical constants characterizing the particular system. It is interesting to note that such pseudobinary systems have an approximately linear relationship between the critical temperature of the terminal binary compounds and the lattice constant of the ternary solid solution over a certain range of concentrations (Fig. 5). This relationship is also confirmed in other similar systems, for example, the systems of V$_3$Sn with Nb$_3$Sn and Ta$_3$Sn [27]. We may thus make an approximate estimate of the critical temperatures in such systems on the basis of a minimum number of experimental points.

Conclusions

We have considered the crystal structure of binary vanadium compounds as a function of the position of the second element in the periodic system. We have analyzed the possibilities of forming compounds with a specific crystal structure in unknown binary systems incorporating vanadium.

By analyzing the critical temperatures of known superconducting binary compounds we have considered the possibilities of new superconducting compounds occurring in binary vanadium systems. It would appear that among the binary vanadium compounds for which the superconducting transformation temperatures are still unknown a high critical temperature can only be expected for V$_3$Al and V$_3$In.

The search for new superconducting compounds is most promising among ternary systems in which compounds of the Cr$_3$Si type are formed. The alloying of binary vanadium compounds of the Cr$_3$Si type with a high critical temperature always leads to a fall in the latter. When continuous solid solutions are formed between superconducting compounds of the Cr$_3$Si type the critical temperature of the ternary solid solutions varies smoothly as a function of composition and may be expressed by empirical functions. In some such pseudobinary systems, in certain regions of concentration there is an almost linear relationship between the critical temperature and the lattice constant of the ternary alloys.

LITERATURE CITED

1. E. M. Savitskii and V. V. Baron, Izv. Akad. Nauk SSSR, Metallurgiya i Gornoe Delo, No. 5, p. 3 (1963).
2. B. T. Matthias, T. H. Geballe, and V. B. Compton, Rev. Mod. Phys., 35(1):1 (1963).
3. B. T. Matthias, Progress in Low-Temperature Physics, Vol. II, Amsterdam (1957).
4. O. Kubaschewski, The Physical Chemistry of Metallic Solutions and Intermetallic Compounds, Vol. 1, No. 2, London (1959).

5.   L. Pauling, The Nature of the Chemical Bond, Cornell University Press (1960).

6.   N. V. Ageev, Nature of the Chemical Bond in Metallic Alloys [in Russian], Izd. AN SSSR (1947).

7.   Yu. M. Golutvin, Heats of Formation and Types of Chemical Bond in Inorganic Crystals [in Russian], Izd. AN SSSR (1962).

8.   D. Laves, in: "Theory of Phases in Alloys" [Russian translation], Metallurgizdat (1961), p. 111.

9.   T. V. Masal'skii, ibid., p. 49.

10.   I. I. Kornilov, The Chemistry of Metallides, Consultants Bureau, New York (1966).

11.   E. Teatum, K. Gschneider, and J. Waber, Los Alamos Scient. Lab. Report 2345 (1960).

12.   Planning the Nomenclature of Inorganic Compounds [in Russian], Izd. AN SSSR (1959).

13.   M. Hansen and K. Anderko, Constitution of Binary Alloys, McGraw-Hill, New York (1958).

14.   R. Kieffer and R. Braun, Vanadin–Niob–Tantal, Berlin (1963).

15.   M. V. Nevitt, Electron Structure and Alloy Chemistry of the Transition Elements, New York (1965), pp; 101–178.

16.   P. I. Kripyakevich, M. A. Tylkina, and E. M. Savitskii, Izv. VUZov, Chernaya Met., No. 1, p. 12 (1960).

17.   A. A. Lena, Metal Progress, No. 66, p. 122 (1954).

18.   P. A. Beck, J. B. Darby, and O. P. Arora, J. Metals, No. 8, p. 148 (1956).

19.   W. Hume-Rothery, Electrons, Atoms, Metals and Alloys. Dover, New York (1962).

20.   W. Buckel, G. Dummer, and W. Gey, Z. Angew. Phys., 14(2):703 (1962).

21.   E. Bucher, F. Neiniger, and J. Müller, Physik der kondensierten Materie, 2(3):210 (1964).

22.   G. F. Hardy and J. K. Hulm, Phys. Rev., 93:1004 (1954).

23.   N. E. Alekseevskii, E. M. Savitskii, V. V. Baron, and Yu. V. Efimov, Dokl. Akad. Nauk SSSR, 145(1):82 (1962).

24.   E. Saur and C. Voepel, Z. Physik, 176:474 (1963).

25.   E. M. Savitskii, V. V. Baron, Yu. V. Efimov, and E. I. Gladyshevskii, Izv. Akad. Nauk SSSR, Neorg. Mat., 1(2):208 (1965).

26.   E. M. Savitskii, V. V. Baron, Yu. V. Efimov, V. R. Krasik, T. V. Vylegzhanina, and E. I. Gladyshevskii, Zh. Neorg. Khim., 9(8):2045 (1964).

27.   E. Rudy, H. Nowotny, F. Benesovsky, R. Kieffer, and A. Neckel, Mh. Chem., 91(1):176 (1960).

SUPERCONDUCTING PROPERTIES OF V—Si
AND V—Ga DIFFUSION COATINGS

Yu. V. Efimov, V. V. Baron, E. M. Savitskii,
and S. N. Sokolov

The effect of diffusion-annealing time and temperature, the amounts of Si and Ga reacting, and other conditions on the phase composition, quality, thickness, and superconducting properties of V—Si and V—Ga coatings obtained by the diffusion of Si or Ga from the vapor phase into a vanadium wire is considered. The optimum conditions for obtaining such coatings are indicated. The best superconducting properties are achieved for single-phase coatings consisting almost entirely of V_3Sn or V_3Ga. The superconducting transformation temperature of V—Si coatings reaches 17°K and that of V—Ga coatings exceeds 14°K. The critical current density is $1 \cdot 10^4$ to $6 \cdot 10^5$ A/cm² at 26 kOe.

The superconducting compounds V_3Ga and V_3Si (Cr_3Si type of structure) have high superconducting transformation temperatures and critical magnetic fields. As regards the temperature of passing into the superconducting state V_3Ga (16.5°K) and V_3Si (17.1°K) are only inferior to Nb_3Sn (18.07°K) and Nb_3Al (17.5°K) [1]. As regards the critical magnetic field V_3Ga exceeds all known superconductors. Extrapolation to 0°K gives a critical field of 350 to 800 kOe for $V_{2.95}Ga$ [2]. An additional proof of the high critical magnetic field for V_3Ga is the manner in which the specific heat varies with temperature in an external magnetic field [3]. The critical magnetic field of V_3Si (150 kOe) is lower than that of Nb_3Sn and V_3Ga [2]. The critical current density of massive samples of V_3Ga, V_3Si and Nb_3Sn in magnetic fields up to 80 kOe is of the same order (about 10^5 A/cm²) [4].

The compounds V_3Si and V_3Ga have a complex crystal structure and are extremely brittle materials. The brittleness of the material interferes with ordinary methods of obtaining wire from these compounds.

During our investigations we learnt of the attempts of E. Sauer and his colleagues [5-7] to obtain wire from V_3Ga and V_3Si by the Kunzler method and by the diffusion of gallium or silicon from the liquid or gaseous phases into vanadium wire. The maximum superconducting transformation temperature of V—Ga and V—Si wire with a superconducting core was 16.4 and 16.9°K respectively.

The maximum critical temperature of diffusion-type V—Ga and V—Si samples, respectively equal to 15.05 and 16.95°K, was achieved after diffusion annealing for 20 h at 1100 and 1000°C respectively. The critical current of samples 0.5 mm in diameter obtained by Sauer et al. in a field of about 25 kOe was 2 to 7 A, which (referred to the thickness of the coating) corresponds to a current density of $1 \cdot 10^4 - 1 \cdot 10^5$ A/cm². The phase composition of the coatings and samples was not studied.

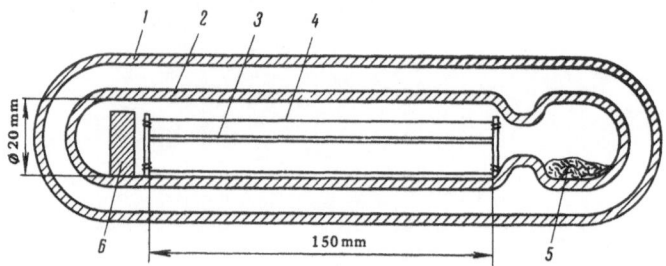

Fig. 1. Ampoule for obtaining diffusion coatings. 1)
Outer quartz ampoule; 2) inner quartz ampoule; 3)
quartz spacer; 4) vanadium wire; 5) silicon or gallium;
6) zirconium getter.

The aim of the present investigation was to study the interactions of vanadium with silicon vapor and gallium vapor and also the superconducting properties of V–Si and V–Ga coatings.

As original materials we took gallium of 99.99% purity, 99.8% silicon, and carbothermal vanadium, the purity of which after refining with cerium or purifying in the electron-beam furnace attained 99.9%.

Vanadium wire 0.2 and 0.5 mm in diameter was prepared by the cold working of cast bars without intermediate annealing. The state of the wire surface (substrate) had a considerable influence on the quality of the diffusion coatings. Good adhesion between the coating and substrate was ensured by the interaction between the latter and the deposited material. The presence of oxides and other contaminations on the surface of the metallic substrate impeded contact between the substrate and coating, prevented diffsuion, and weakened the substrate-coating adhesion. The graphite lubricant was removed by drawing the vanadium wire in the dry state for the last few passes. After mechanical cleaning the wire was degreased in a 10% aqueous solution of alkali at 50°, washed in hot and cold water, and washed in alcohol. For final cleaning of the surface before coating, the vanadium wire was annealed in a vacuum of 10^{-5} mm Hg for 1 h at 1000°.

The deposition of the coating and the heat treatment of the wire samples 0.2 and 0.5 mm in diameter and 150 to 500 mm long were carried out in evacuated quartz ampoules in muffle and Silit resistance furnaces, the temperature of these being regulated with a Pt-PtRh thermocouple.

The cleaned vanadium wire, fixed to a quartz spacer, was placed in the quartz ampoule, the appropriate amount of silicon or gallium being held in a separate container (Fig. 1). In order to remove traces of air a zirconium getter was enclosed in the ampoule on sealing; this had little effect on the composition of the coatings. In order to obtain a coating from the gas phase the vanadium wire and the corresponding amount of silicon or gallium were kept in double evacuated quartz ampoules for 1 to 100 h at 700 to 1300°. Complex heat treatment, including various modes of diffusion annealing in order to obtain the coatings and subsequent aging of the coated samples at various temperatures in vacuo without silicon or gallium, was also applied.

In order to obtain coatings from the liquid phase, the ampoules with the vanadium wires were completely filled with gallium. The annealing period in this case was much shorter.

The phase composition of the coatings was studied by x-ray analysis. The x-ray pictures were taken directly from the wire in an RKD camera in copper radiation with a nickel filter. For calculation purposes we used standard x-ray photographs of V-Ga and V–Si alloys obtained earlier in conjunction with colleagues from the I. Franko L'vov State University. The phases formed in the coatings gave several different values of interplane spacing; this was probably due to their nonequilibrium nature.

We also made a microscope study of the cross sections and surface state of the resultant samples; with the help of an object micrometer we estimated the thickness of the coatings and refined this by reference to the weight of the samples before and after diffusion annealing. The microsections were etched electrolytically in a 5 to 10% aqueous solution of HCl. The adhesive strength of the coatings relative to the base was estimated qualitatively from the bending of the samples.

The critical current was measured in samples 0.2 to 0.5 mm in diameter and 100 to 140 mm long at 4.2°K, three or four times for each value of the transverse magnetic field (up to 26.6 kOe). The superconducting transformation temperature of the samples was determined in a specially-developed apparatus made by N. D. Kozlova of the Institute of Metals.

The interaction of silicon vapor with the vanadium and the formation of V−Si coatings on the vanadium wire started at 750°. Gallium evaporated considerably on heating to 700°. However, at these temperatures the amounts of silicon and gallium vapor formed were insufficient to yield coatings in brief holding periods (up to 20 h). After holding at 700° the x-ray diffraction pictures of the vanadium wire showed hardly any more than the lines of pure vanadium. For higher temperatures the silicon and gallium vapor interacted vigorously with the vanadium, forming uniformly-thick coatings of identical composition (for the same heat treatment). The thickness of the coatings fluctuated with annealing time and temperature and with the original amount of silicon or gallium, referred to unit surface of the vanadium wire; the optimum thickness was about $1\,\mu$ for V−Si and $2.5\,\mu$ for V−Ga coatings. The microstructure of the coatings was complex. Individual layers could not always be resolved in multilayer coatings. In order to recognize these clearly, color-etching was required. After coating in vacuum the resultant wire was fairly ductile. On bending, the coatings obtained under optimum conditions neither peeled nor ruptured.

On diffusion-annealing at temperatures up to 800° for 20 to 100 h with any amounts of silicon or up to 900° for short periods with large amounts of silicon (over 1 g), the compound VSi_2 was formed. A layer of silicon which had failed to react completely was formed on the surface of the samples, particularly after annealing with large quantities of the evaporating material. On raising the annealing temperature, however, and annealing for 5 to 100 h, the silicon reacted completely with the vanadium for any quantities (up to 8 g). The compound VSi_2 remained in the coatings after annealing at all temperatures up to 1200° for amounts of silicon greater than 0.3 g. After annealing above 1100° the amounts of VSi_2 remaining in the coatings were quite small.

The principal phase in the coatings after annealing at 1100 to 1200° was V_5Si_3. The amount of this compound, like that of VSi_2, depended on the annealing time and the amount of silicon interacting. Additional annealing at the same temperatures (without silicon) increased the amount of V_5Si_3 in the coatings of samples initially treated at 1000 to 1200°.

A small amount of V_3Si appeared in the coating even after annealing at 1000° (20 h, 0.1 to 1 g of Si). On raising the annealing temperature above 1100°, particularly for amounts of silicon between 0.1 and 0.3 g, the quantity of V_3Si in the coating increased. After annealing above 1200° the coating consisted mainly of V_3Si (for holding periods of at least 2 h and silicon weights up to 1.5 g).

When gallium vapor interacted with vanadium the picture was very complicated. However, the results obtained enabled us to choose a suitable set of conditions for obtaining V−Ga coatings only containing V_3Ga.

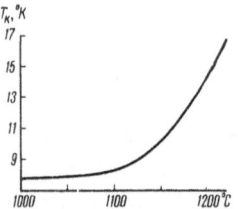

Fig. 2. Superconducting transformation temperature of V—Si diffusion coatings as a function of annealing temperature.

For large amounts of VSi_2 in the coatings, the superconducting transformation temperature of V—Si coatings was 7.7°K, and that of the compound V_5Si_3 7.9°K. The appearance of superconductivity in the compound VSi_2 (nonsuperconducting in the ordinary state) and the rise in the critical temperature of the compound V_5Si_3 were evidently due to the film effect [8, 9].

With increasing diffusion-annealing temperature the T_K of the V—Si coatings increased (Fig. 2). The single-phase coating obtained at 1220° and consisting solely of the compound V_3Si passed into the superconducting state at 17.0°K.

The period of diffusion annealing had a considerable effect on the critical temperature. For example, on increasing the annealing period at 1220° from 1 to 5 h (0.5 g of Si) the T_K of the coatings increased from 14 to 17°K. An increase in annealing period promoted the formation of more equilibrium phases in the coatings.

Samples with vanadium—silicon coatings obtained at annealing temperatures of up to 1200° passed into the superconducting state over a finite range of temperatures (diffuse transformation), owing to the presence of several phases in the coating. The transformation curves showed several characteristic steps, the number of these corresponding to the number of phases in the coatings. A second possible reason for the diffuseness of the transformation was a certain departure from equilibrium in the phases formed in the coatings, attributable to the actual diffusion method of obtaining the samples. After annealing at temperatures above 1220°, when the coating consisted of V_3Si only, there was a sharp transition into the superconducting state (Fig. 3).

The character of the transformation curve and the critical temperature of the coatings were considerably affected by the weight of silicon taken, referred to unit surface of the vanadium wire (Fig. 4). For each diffusion-annealing temperature a specific optimum amount of silicon (q) per unit surface of the vanadium wire had to be taken in order to achieve the maximum T_K and a sharp transformation curve. At 1220° the maximum T_K (17.0°K) was achieved with q = 0.054 g/mm^2, corresponding to a coating thickness of 1μ. On increasing or reducing the amount of silicon the critical temperature of the resultant coating fell or rose respectively.

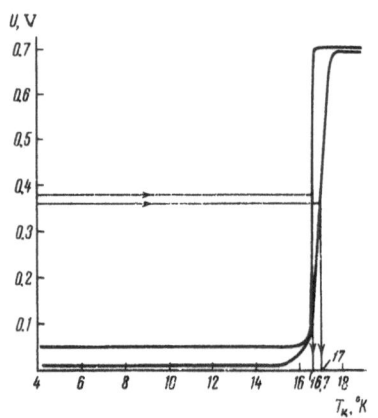

Fig. 3. Transformation curves of V—Si diffusion coatings obtained at temperatures of over 1220°.

Vanadium—gallium coatings mainly containing the compound V_2Ga_5 had a very low T_K (7.3°K); in the presence of the δ phase the T_K rose to 7.7 or 8.0°K. The superconducting transformation temperature of coatings containing V_3Ga reached 8 to 14°K, depending on the conditions of production. The critical temperature of single-phase coatings consisting of V_3Ga was higher than 14°K. The superconducting properties of V—Ga coatings were more severely affected not only by the amount of gallium interacting and annealing time and temperature, but also by other conditions of the process.

The main difficulty in studying the critical current of the samples obtained lay in ensuring reliable contact between the sample and the measuring leads of the apparatus. Unreliable soldering or the rupture

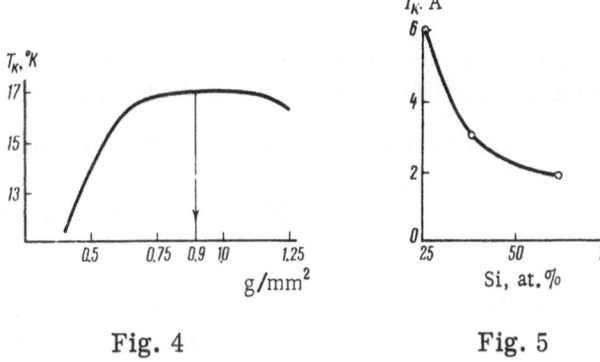

Fig. 4   Fig. 5

Fig. 4. Superconducting transformation temperature of V—Si diffusion coatings obtained at 1220° as a function of the amount of interacting silicon per unit surface of the vanadium wire.

Fig. 5. Critical current of V—Si diffusion coatings as a function of silicon content.

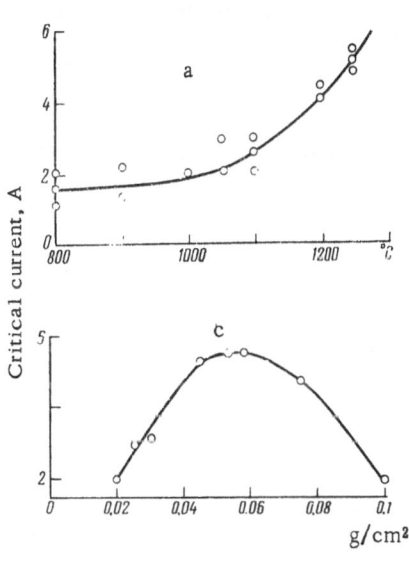

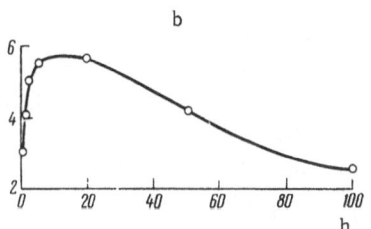

Fig. 6. Critical current of V—Si coatings as a function of the temperature of diffusion annealing (a), the period of annealing at 1200° (b), and the amount of interacting silicon, referred to unit area of the vanadium wire (c).

of the coatings on welding impeded measurement of the critical currents. The best results were obtained with rubbing contacts.

The effect of silicon content and hence of phase composition of the coatings on the critical current of V—Si coatings is shown in Fig. 5.

The effect of annealing time and temperature and also the amount of evaporating silicon (per unit surface of the vanadium wire) on the critical current at 26.6 kOe for V—Si coatings obtained by diffusion annealing at temperatures above 1200° is shown in Fig. 6. The critical current of single-phase vanadium—gallium coatings consisting of the compound V_3Ga at 26.6 kOe is over 6 A. The critical current density referred to unit area of the V—Si and V—Ga coatings over the cross section equals $1 \cdot 10^4 - 6 \cdot 10^5$ A/cm^2.

Conclusions

As a result of the foregoing investigations we have established the conditions for obtaining V–Si and V–Ga superconducting diffusion coatings on a vanadium wire.

We studied the effect of annealing time and temperature, the amount of silicon or gallium entering into the reaction (referred to unit surface of the vanadium wire), and other conditions of the process on the phase composition, quality, thickness, and superconducting properties of V–Si and V–Ga diffusion coatings.

The best superconducting properties are achieved in single-phase coatings comprising only V_3Si or V_3Ga. We have determined the optimum conditions for obtaining such coatings.

The superconducting transformation temperature of V–Si diffusion coatings obtained under optimum conditions reaches 17°K, and that of V–Ga coatings over 14°K. The critical current density of the diffusion coatings is $1 \cdot 10^4 - 6 \cdot 10^5$ A/cm^2 at H = 26.6 kOe.

LITERATURE CITED

1. E. M. Savitskii and V. V. Baron, Izv. Akad. Nauk SSSR, Metallurgiya i Gornoe Delo, No. 5, p. 3 (1963).
2. J. H. Wernick, E. J. Morin, F. S. L. Hsu, D. Dorsi, J. R. Maita, and J. E. Kunzler, Proc. Internat. Conf. High Magnetic Fields, Cambridge (1961), p. 609.
3. W. E. Blumberg, J. Elsinger, V. Jallarino, and B. J. Matthias, Phys. Rev. Lett., 5:149 (1960).
4. J. E. Kunzler, E. Buehler, F. S. L. Hsu, and J. H. Wernick, Phys. Rev. Lett., 6:89 (1961).
5. D. Koch, G. Otto, and E. Saur, Phys. Lett., 4(5):1 (1963).
6. I. Babiskin, R. P. G. Siebemann, G. Otto, and E. Saur, Z. Phys., 180(5):483 (1964).
7. C. Muller, G. Otto, and E. Saur, Z. für Naturforsch., 19a(5):539 (1964).
8. M. Tanenbaum and W. V. Wright, J. Metals, 14(5):367 (1962).
9. Electronics, 34(29):88 (1961).

STRUCTURE AND SUPERCONDUCTING PROPERTIES
OF V–Ga ALLOYS

Yu. V. Efimov, V. V. Baron, and E. M. Savitskii

The phase diagram of the V–Ga system is refined by microstructural and thermal analysis, x-ray diffraction, measurements of microhardness and critical superconducting temperature, and also the determination of the chemical composition of certain phases by x-ray spectral microanalysis. The results are compared with published data. The effect of chemical composition and heat treatment on the superconducting temperature of the β phase (Cr_3Si structure) is studied. The transformation temperature of the β phase is affected by the degree of ordering of its crystal structure and the degree of uniformity of the alloys. The maximum transformation temperature occurs for 26 at.% Ga. Gallium reduces the critical superconducting temperature of vanadium.

The V–Ga system has attracted the attention of research workers by reason of the high superconducting properties of the compound V_3Ga ($T_K = 16.5°K$ [1-3], $H_K > 350$ kOe [4]).

We published the first approximate phase diagram of the V–Ga system in [5]. Apart from earlier-known compounds, namely, V_3Ga with the Cr_3Si structure (a = 4.816 Å [1]) and V_2Ga_5 with the Mn_2Hg_5 structure (a = 8.96 Å, c/a = 0.300) [6], the system contained three new compounds with roughly 40, 50, and 85 at.% of Ga. We found that gallium was quite soluble in vanadium and that there was a psuedoeutectic equilibrium on the gallium side.

Recently three more versions of the V–Ga phase diagram have been published (Fig. 1) [7-9]. There have also been a number of papers devoted to particular phases of this system [10-14]. Published data regarding the structure of individual regions of the V–Ga phase diagram, the crystal structures of individual phases, and the conditions for the formation of these are to a certain extent contradictory. Data relating to the influence of the composition of the alloys on the superconducting transformation temperature of V_3Ga also differ [8, 13].

The aim of the present investigation was to refine the phase equilibria in the V–Ga system and to study the effect of the composition and heat treatment of the alloys on the superconducting transformation temperature of V_3Ga.

In order to prepare the alloys we mainly used carbothermal vanadium refined with cerium (99.9 wt.% V). Some parts of the phase diagram were studied with alloys prepared from electrolytic vanadium previously remelted in an electron-beam furnace (over 99.9 wt.% V). The original gallium contained less than 0.01 wt.% of impurities. The method of preparation, treatment, and examination of the alloys was essentially the same as that employed in [5]. We found a certain reduction in the amount of gallium in alloys, attributable to evaporation during high-temperature heat treatment; for this reason experimental samples of new alloys were subjected to additional chemical analysis after heat treatment, particularly when this was conducted at high temperatures.

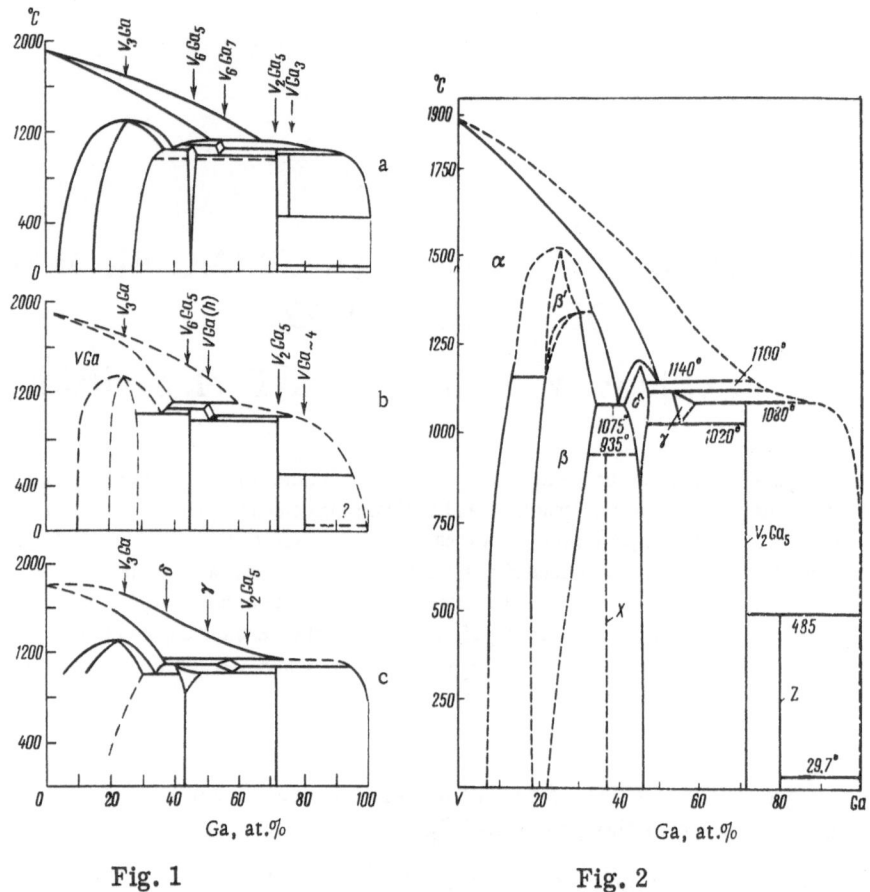

Fig. 1   Fig. 2

Fig. 1. Versions of the V—Ga phase diagram proposed by a) V. M. Pan [7], b) H. Meissner and K. Schubert [9], and c) Van Vucht et al. [8].

Fig. 2. Phase diagram of the V—Ga system [15].

The alloys were studied by microstructural and thermal analysis and x-ray diffraction. The microhardness of individual phases in the alloys was also measured. In order to determine the chemical composition of some phases more accurately, some of the alloys were also subjected to x-ray microanalysis.*

The superconducting transformation temperature was measured by a magnetic method on monolithic samples or samples pressed from powder, 7 to 10 mm long and 2 to 2.5 mm in diameter. The accuracy of the critical temperature measurement was ± 0.2°K.

The phase diagram thus plotted is shown in Fig. 2 [15].

Solid solutions based on both metals are formed in the system.

The V-base solid solution has a wide range of homogeneity. The maximum solubility of Ga in V reaches 50.4 at.%. Our own maximum solubility agrees with that of V. M. Pan [7]. The authors of [8, 9] give lower solubilities (about 40 at.% Ga).

---

* The authors are extremely grateful to Master of Physicomathematical Sciences P. I. Kripya-kevich (Franko L'vov State University) for the x-ray analysis of the alloys and Master of Physicomathematical Sciences V. A. Batyrev (Baikov Institute of Metallurgy) for the x-ray spectral microanalysis of the samples.

A study of alloys with higher purities confirmed our value for the solubility of gallium in vanadium at 800°, about 10 at.% Ga [5]. On further reducing the temperature the solubility of gallium in vanadium changes little (8.5 at.% at 600°, about 8 at.% at 400°).

The V-base solid solution with the limiting gallium content, in our opinion, decomposes at 1140° into the δ phase and a liquid containing about 70 at.% Ga.

A very low solubility of vanadium in gallium (0.03 to 0.05 at.%) was established by x-ray diffraction and x-ray spectral microanalysis. On the gallium side the system contains in fact a degenerate eutectic. The eutectic point lies at 0.11 at.% V. The difference in the melting point of pure gallium (29.75° [16]) and the eutectic is less than 0.1° (of the order of 0.05° [7]).

The compound V_3Ga is formed from the solid solution based on vanadium, according to our own data (as distinct from the results of others [7-9]), at 1525°. In our opinion, the compound V_3Ga undergoes a polymorphic transformation at 1155 to 1340° (depending on the composition of the alloys within the range of homogeneity of the phase).

The high-temperature form of the compound (β' phase) is stable at 1155 to 1525° over the range 16 to 37 at.% Ga. The maximum width of the range homogeneity of the β' phase is evidently achieved at 1340° (31 to 32 at.% Ga). According to [13] the high-temperature form (1100 to 1500°, 22 to 33 at.% Ga) has an hcp lattice (a= 4.1, c = 5.3 Å at 25 at.% Ga). According to our data, the β' phase only differs from the V-base α solid solution in respect to the partial ordering in the positions of the V and Ga atoms in the structure, as indicated by the appearance of extra lines on the x-ray diffraction pictures of the alloys. We consider that the ordered structure belongs to the CsCl type. In view of the fact that the range of homogeneity of the β' phase fails to reach the composition VGa, the ordering in the structure cannot be complete; it should obey the formula $V(V_{0.5}, Ga_{0.5})$. The brackets enclose the symbols of the components with a disordered distribution of the atoms.

The low-temperature form of the compound V_3Ga (the β phase) has a cubic structure of the Cr_3Si type. The maximum width of the range of homogeneity of this phase (22 to 35 at.% Ga) occurs at 1075°.

The delta phase has the composition V_6Ga_5 [10]. In our view the δ phase is not formed by a peritectic reaction [7-9] but in the same way as the compound V_3Ga, from the vanadium-base solid solution at 1195° and about 45.5 at.% Ga. At high temperatures the δ phase has a small range of homogeneity, reaching its maximum width (42.5 to 49 at.% Ga) at about 1100°. This phase has a hexagonal lattice with a = 8.51, c = 5.19 Å, c/a = 0.61. The values obtained almost coincide with those of [10]. The strong lines on the x-ray picture of the δ phase may also be indexed on the basis of a hexagonal structure with a = 4.91 Å = $a_0 \cdot 2\sqrt{2/3}$, where a_0 is the lattice constant of the cubic substructure of the α-Fe type [5]. The lattice constants found in [10] and supported by our own data are also related to the period a_0: a = $a_0 \cdot 2\sqrt{2}$, c = $a_0\sqrt{3}$. Hence the structure of the δ phase should be related to that of α-Fe.

Between the δ phase and the compound V_3Ga a eutectic is formed at 1075° and 40.5 at.% Ga.

In the region between the β and δ phases a peritectic reaction at about 935° produces yet another phase with a hexagonal lattice (a = 7.356, c = 4.971 Å, c/a = 0.676). We were quite unable to produce this phase (the x phase) in pure form. The lattice parameters of the x phase are characteristic of a structure of the Mn_5Si_3 type and similar to those obtained for the ternary phase $V_5Ga_3O_x$ [9, 14]. It is possible that this x phase was in fact the compound $V_5Ga_3O_x$. However, it is not impossible that there might be a corresponding binary compound V_5Ga_3 formed from V_3Ga and the δ phase by a peritectoid reaction at about 935°. At any rate, the x phase was also observed in alloys of this range quite free from oxygen (under 0.1 wt.% O). It is also known that this phase is not stabilized by nitrogen [11, 12] or carbon [15].

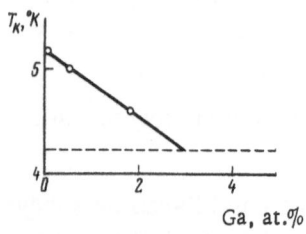

Fig. 3. Superconducting transformation temperature of vanadium as a function of the amount of gallium added.

The gamma phase (V_6Ga_7), established earlier in [10], has a bcc lattice and is formed at 1110° as a result of the peritectic interaction of the δ phase with the liquid containing about 80 at.% Ga. Quenching alloys with 50 to 60 at.% Ga from 1100° shows that the composition of the γ phase formed is close to 54 at.% Ga. This phase is only stable at high temperatures; at 1020° and 56 at.% Ga it decomposes by a eutectoid transformation into the δ phase (48 at.% Ga) and V_2Ga_5. The range of homogeneity of the γ phase widens in the sense of increasing the gallium content (slight change in the lattice constant, a = 9.18-9.20 Å) and at the limit reaches 59 at.% Ga at 1080°. The x-ray diffraction picture of the γ phase is generally reminiscent of compounds of the γ-brass type (e.g., Cu_5Zn_8). The number of atoms in the unit cell of the γ phase in the alloy containing 57 at.% Ga equals 54 (ρ = 6.84) and in the alloy containing 58 at.% Ga 52 (ρ = 6.77 g/cm³). Thus the γ phase evidently belongs to the class of phases with a variable number of atoms in the unit cell, and its structure is of a transitional nature between the nondefective superstructure of the α type and the defective structure of the γ-brass variety.

Regarding the compound V_2Ga_5 there is no particular disagreement in the literature.

The Ga-rich z phase is formed by a peritectic reaction at 485° [5]; it contains about 80 to 85 at.% Ga [5, 9, 15] and plainly has a distorted structure of the $NiHg_4$ type [9].

We failed to establish the existence of the compound VGa_3 mentioned in [7].

Measurements of the superconducting-transformation temperature of V–Ga alloys showed that only β phase with the cubic Cr_3Si type of structure and the low-alloy V-base solid solution possessed superconductivity. The remaining V–Ga alloys were not superconducting, at any rate above 4.2°K.

The superconducting transformation temperature of electrolytic vanadium remelted in an electron-beam furnace is 5.3°K. On dissolving Ga the critical temperature of vanadium falls and at about 3 at.% Ga lies below 4.2°K.(Fig. 3). Solid solutions of gallium and vanadium containing over 3 at.% Ga are not superconducting at temperatures above 4.2°K.

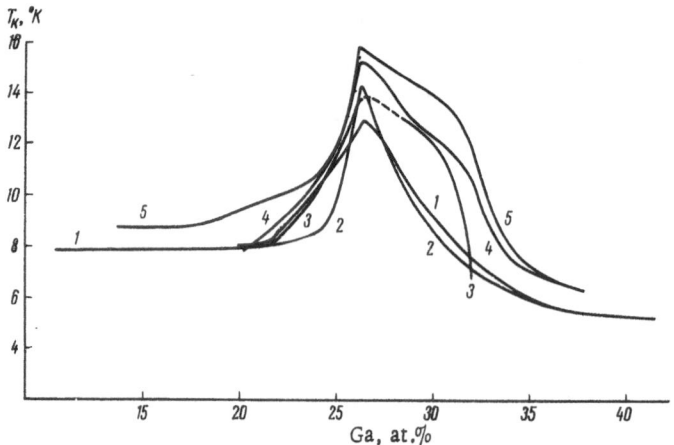

Fig. 4. Superconducting transformation temperature of the β phase (structure of the Cr_3Si type) as a function of the chemical composition of the alloys and the heat treatment applied. 1) Cast; annealed at 2) 800°; 3) 900°; 4) 1050°; 5) 1150°.

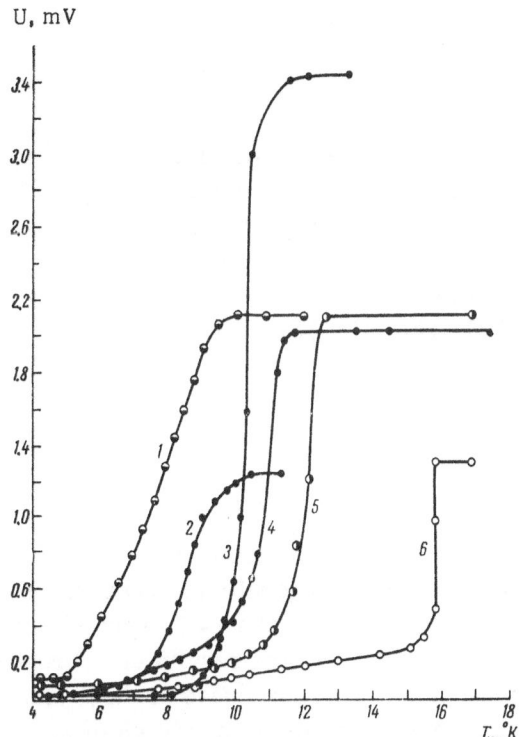

U, mV

Fig. 5. Transformation curves of V–Ga alloys containing the β phase. Gallium content in at.%: 1) 31.8 (cast); 2) 21.5 (cast); 3) 26.3 (annealed at 800°); 4) 31.8 (annealed at (1050°); 5) 31.8 (annealed at 1150°); 6) 25.9 (annealed at 1150).

Depending on the conditions of production and the heat treatment, the superconducting β phase is formed in alloys containing from 8 to 46 at.% Ga (see Fig. 2).

The superconducting transformation temperature of these alloys is shown as a function of chemical composition and heat treatment in Fig. 4. The value of the critical temperature of the β phase depends on its chemical composition. In two-phase samples lying close to the range of homogeneity of the β phase the critical temperature varies little. In equilibrium with the V-base solid solution the critical temperature of the β phase is 7.9 to 8.8°K. The similar values of the critical temperatures of these alloys after different kinds of heat treatment may be explained by the small variation in the Ga content of the V_3Ga with varying temperature. On passing into the range of homogeneity the critical temperature starts rising, reaching a maximum at about 26 at.% Ga. The displacement of the critical-temperature maximum from the compound of stoichiometric composition in the direction of an alloy with a greater gallium content agrees with the analogous displacement in the composition corresponding to complete ordering of the β phase [13]. On further raising the gallium content there is a fall in critical temperature (slow within the range of homogeneity and rapid on passing into the two-phase region). In the two-phase region lying close to the range of homogeneity of the β phase on the gallium side, the β phase has a lower superconducting transformation temperature (5.5 to 6.5°K) than that of alloys in the two-phase region on the vanadium side.

In the cast state the β phase has a comparatively low transformation temperature (up to 13.1°K). The transformation curves of these alloys are diffuse (1 and 2 in Fig. 5). The comparatively low critical temperatures and the diffuse superconducting transformation of the cast alloys are evidently explained by their inhomogeneity. Apart from the inhomogeneity, different parts of the β phase in the cast alloys have certain differences in composition owing to the occurrence of solid-state transformations in the course of cooling. All this gives rise to variations in the critical temperatures of these regions and hence to a diffuse transformation.

For alloys homogenized over a long period we obtain sharper transformation curves (3 to 6 in Fig. 5). With increasing homogenization temperature these curves become sharper, and the superconducting transformation temperature rises (Fig. 6). The highest critical temperatures occur for alloys quenched after homogenization from 1150°. Actually our own maximum temperature, equal to 15.8°K for the alloy containing 26 at.% Ga, is slightly lower than that mentioned in the literature, 16.5°K [1-3]. The rise in the critical temperature of alloys containing the β phase on increasing the homogenization temperature is evidently associated with the achievement of a more equilibrium state and the occurrence of ordering processes. With increasing homogenization temperature, the mobility of the atoms becomes greater, and diffusion processes and the rearrangement of the crystal lattice occur more rapidly. A rise in the critical temperature on ordering was noted, for example, in alloys of the Nb_3Sn–Nb_3Ga–Nb_3Sb system [17].

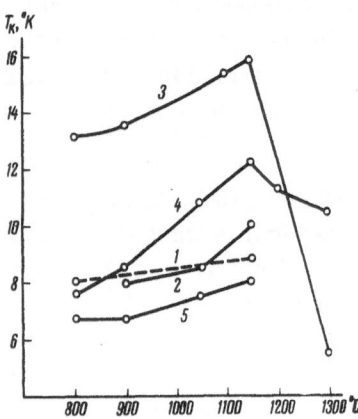

Fig. 6. Critical temperature of alloys containing the β phase as a function of the homogenization temperature Gallium content, at.%: 1) 17.5; 2) 21.4; 3) 25.9; 4) 31.8; 5) 33.9.

It is interesting to note that measurements of the transformation temperature may be used in the phase analysis of alloys and in refining the ranges of homogeneity of superconducting phases. The boundaries of the range of homogeneity of the β phase in the V–Ga system, obtained by measuring the critical temperature, are in full accord with the results of microstructural and x-ray analysis and also with the manner in which the microhardness of the β phase varies. Considering the high sensitivity of the magnetic method to small quantities of superconducting phase (see p. 191 of this collection), we may consider the measurement of the critical temperature as a very promising method of physicochemical analysis.

Conclusions

We have refined the phase diagram of the V–Ga system by microstructural and thermal analysis, x-ray diffraction, measurements of microhardness and superconducting-transformation temperature, and also the x-ray microanalytical determination of the chemical composition of various phases.

We have compared the results obtained with published data.

We have studied the effect of chemical composition and heat treatment on the superconducting-transformation temperature of the β phase. We have found that the maximum critical temperature is that of the β phase containing 26 at.% Ga. We consider that the degree of ordering in the crystal structure of the β phase and the closeness of the alloys to the equilibrium state have a considerable effect on the transformation temperature.

We have found that gallium reduces the critical temperature of pure vanadium. A V-base solid solution containing over 3 at.% Ga is not a superconductor at temperatures above 4.2°K.

The compound V_2Ga_5 and the β', x, δ, γ, and z phases also show no superconductivity above 4.2°K.

LITERATURE CITED

1.   R. A. Wood, V. B. Compton, B. T. Matthias, and E. Corenzwit, Acta Cryst., 11:9 (1958).
2.   G. F. Hardy and J. K. Hulm, Phys. Rev., 89:884 (1953).
3.   E. A. Wood, V. B. Compton, B. T. Matthias, and E. C. Corenzwit, Acta Cryst., 11:604 (1958).
4.   J. H. Wernick, F. J. Morin, F. S. L. Hsu, D. Dorsi, J. R. Maita, and J. E. Kunzler, High Magnetic Fields, Technology Press, Cambridge, Mass. (1962), p. 609.
5.   E. M. Savitskii, P. I. Kripyakevich, V. V. Baron, and Yu. V. Efimov, Zh. Neorg. Khim., 9(5):1155 (1964).

6. K. Schubert, H. G. Meissner, W. Rossteutscher, and E. Stolz, Naturwiss., 49(3):57 (1962).

7. V. M. Pan, Collection "Structure of Metallic Alloys" [in Russian], Izd. AN Ukr.SSR, Kiev, p. 120.

8. J. H. N. Van Vucht, H. A. C. M. Bruning, H. C. Dunkersloot, and A. H. Gomes de Mesquita, Philips Res. Reports, 19:407 (1964).

9. H. G. Meissner and K. Schubert, Z. Metallkunde, 56(7):475 (1965).

10. J. H. N. Van Vucht, H. A. C. M. Bruning, and H. C. Donkersloot, Phys. Lett., 47(5):297 (1963).

11. W. Jeitschko, H. Nowotny, and F. Benesovsky, Mh. Chemie, 95(1):156 (1964).

12. W. Jeitschko, H. Nowotny, and F. Benesovsky, Mh. Chemie, 95(4-5):1212 (1964).

13. H. J. Levinstein, J. H. Wernick, and C. D. Capio, J. Phys. Chem. Solids, 26(7):1111 (1965).

14. K. Schubert, K. Frank, R. Gohle, A. Maldonado, H. G. Meissner, A. Ramen, and W. Rossteutscher, Naturwiss., 50:41 (1963).

15. E. M. Savitskii, P. I. Kripyakevich, V. V. Baron, and Yu. V. Efimov, Izv. Akad. Nauk SSSR, Neorg. Mat., 3(1):45 (1967).

16. M. A. Filyand and E. I. Semenova, Properties of Rare Elements [in Russian], Izd. "Metallurgiya" (1964).

17. N. E. Alekseevskii, N. V. Ageev, and V. F. Shamrai, Izv. Akad. Nauk SSSR, Neorg. Mat., 2(12):2156 (1966).

PRODUCTION AND SUPERCONDUCTING PROPERTIES OF DIFFUSION COATINGS OF INDIUM, CADMIUM, LEAD, THALLIUM, BISMUTH, AND ZINC ON VANADIUM

Yu. V. Efimov, V. V. Baron, and E. M. Savitskii

The interaction of In, Cd, Pb, Tl, Bi, and Zn vapor with vanadium wire is studied after holding isothermally for 1 to 20 h at 1000 to 1200° in evacuated quartz ampoules. X-ray analysis of the resultant binary diffusion coatings reveals the formation of V_3In (a = 5.28-5.56 Å), V_3Cd (a = 4.92-4.95 Å), V_3Pb (a = 4.87 Å), V_3Tl (a = 5.21-5.25 Å), and V_3Bi (a = 4.72 Å), with the Cr_3Si structure. The existence of a compound V_3Zn with an analogous structure appears quite likely. The x-ray diffraction pictures of the coatings also show the lines of vanadium (the base) and those of a third phase, the crystal structure of which remains unsolved. The critical superconducting temperature of the compound V_3Zn obtained by the diffusion method is 13.9°K. The rest of the compounds obtained are not superconducting above 4.2°K.

The development of new superconducting materials with high parameters is one of the main problems in the metallography, physical chemistry, and physics of superconductors [1].

The highest critical magnetic fields and superconducting transformation temperatures occur for compounds with the cubic Cr_3Si type of structure [1, 2]. This is attributable to the particular configuration of energy bands associated with this structure (narrow d band) and the high electron density of states on the Fermi surface [3, 4].

The superconducting vanadium compounds with the Cr_3Si structure known at the present time are shown in Table 1.

The formation of compounds is ultimately due to the arrangement of the interacting elements in the periodic system, i.e., the difference in the structure of their electron shells. Analysis of the types of crystal structure and the conditions for the formation of compounds in binary vanadium systems shows that compounds with the Cr_3Si structure are capable of being formed in alloys of vanadium with nontransition metals of Groups II to V in the periodic system [7].

The aim of the present investigation was to study the interaction of vanadium with certain nontransition metals of Groups II to V and also to examine the phase composition and critical temperature of the corresponding diffusion coatings.

TABLE 1. Binary Vanadium Compounds with the Cr_3Si Structure Studied for Superconductivity [2, 5, 6]

Compound	T_K, °K	Compound	T_K, °K
V_3Si	17.1	V_3Sb	0.80
V_3Ga	16.8	V_3Au	0.74
V_3Sn	7.00	V_3Rh	0.38
V_3Ge	6.01	V_3As	< 1.02 *
V_3Pt	2.83	V_3Co	< 0.35 *
$(V_{2.67}, Ir_{0.33})$ Ir	1.39	V_3Ir	< 0.35 *

* No superconductivity found above the temperature indicated.

154

The original materials were vanadium, cadmium, zinc, indium, thallium, lead, and bismuth more than 99.9 wt.% pure. The interaction between the vapor of these nontransition metals and the vanadium (taken in the form of wire 0.2 mm in diameter) was carried out in a closed space (evacuated quartz ampoules) while holding isothermally at 1000 to 1200° for 1 to 20 h. The weight of material for evaporation was 0.5 g per 1 cm² of the surface of the vanadium wire.

The phase composition of the resultant coatings was established by x-ray analysis. The x-ray diffraction pictures were taken in RKU and RKD cameras 86 and 57.3 mm in diameter in vanadium or copper radiation. The samples were diffusion-coated wires (0.2 mm diameter). The crystal lattice parameters were determined to an accuracy of ± 0.01 A.

The superconducting transformation temperature was determined by the magnetic method from the change in the magnetic susceptibility of wire samples at temperatures above 4.2°K.

The critical temperature was determined by reference to the middle point of the transformation curve.

X-ray analysis of the resultant coatings showed that the vapors of all the nontransition metals studied interacted readily with vanadium at 1000 to 1200°, and formed the corresponding binary coatings on the latter (V–Zn, V–Cd, etc.) in periods of over 5 h.

All the coatings were multiphased. Apart from the lines of vanadium (the base), the x-ray pictures all showed the lines of phases with a cubic Cr_3Si-type structure (V_3In, V_3Cd, V_3Pb, V_3Tl, V_3Bi, and possibly V_3Zn), and also the lines of at least three other phases, the crystal structure of these latter not being solved.

The compounds V_3Cd (a = 4.943 Å) and V_3Pb (a = 4.937 Å) are well known from the literature. These compounds were first observed in alloys prepared by the metalloceramic method [8]. The production of V_3In was mentioned in [9]. The compounds V_3Tl and V_3Bi with the cubic Cr_3Si structure were obtained for the first time in the present investigation.

We consider that the formation of a compound V_3Zn is quite possible, although its existence requires further proof. The x-ray pictures of the V–Zn coatings showed two or three very weak and diffuse lines which might have belonged to this phase. Some of the lines may overlap lines of other phases in the V–Zn system [2, 10, 11]. We were unable to obtain sharper lines under the conditions of diffusion annealing available.

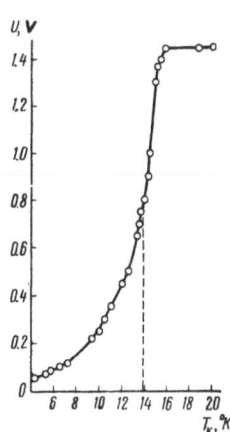

Fig. 1. Transformation curve of a diffusion coating containing V_3In.

The lattice constants a of the Cr_3Si-type compounds obtained by the diffusion method are 4.92-4.95 Å for V_3Cd, 4.87 A for V_3Pb, 5.28-5.56 Å for V_3In, 5.21-5.25 Å for V_3Tl and 4.72 Å for V_3Bi. The variation in the lattice constant of V_3Cd, V_3In, and V_3Tl (as a function of production conditions) indicates the existence of a range of homogeneity. The compounds V_3Pb and V_3Bi evidently have a constant composition with a very narrow range of homogeneity.

The lattice constant a of the wire base (vanadium) with V–Zn coatings rose from 3.028 to 3.05 Å. In the samples with other coatings the lattice constant of vanadium changed much less. Evidently the solubility of zinc in vanadium is quite high and that of the other metals studied insignificant.

On studying the critical temperature of the coatings it was found that V_3In passed into the superconducting state at 13.9°K (see graph, Fig. 1). The slightly diffuse temperature range of the transition is due to a certain variation in the composition of the V_3In through the thickness of the coating, attributable to the production method itself. The diffusion coatings of other metals show no superconductivity at temperatures above 4.2°K. The critical temperature of vanadium falls slightly on alloying with these metals (on obtaining coatings by the diffusion method).

In compounds of the Cr_3Si type the electron concentrations favorable toward high critical temperatures eqúal 4.5 and 6.5 electrons/atom approximately [12]. Of the compounds obtained the one with the most favorable electron concentration is V_3In, and it is this which has a high transformation temperature. The electron concentrations of V_3Bi, V_3Pb and V_3Cd are less favorable. It is thus quite understandable why these compounds are not superconducting at temperatures above 4.2°K. The compounds V_3Ti and V_3In have the same electron concentration. However, thallium is an element of the third long period. On replacing the B component in a compound of the Cr_3Si type by an element of the same group in the periodic system, an increase in the number of the period leads to a fall in the critical temperature.

Conclusions

Coatings of Cd, In, Tl, Pb, Bi, and Zn containing the compounds V_3Cd, V_3In, V_3Tl, V_3Pb and V_3Bi with the cubic Cr_3Si type of structure were produced on vanadium by diffusion from the gas phase at 1000 to 1200°. This was the first time that V_3Tl and V_3Bi had been observed. The existence of a compound V_3Zn with an analogous structure also appeared quite probable.

The superconducting transformation temperature of the V_3In obtained by the diffusion method was 13.9°K. None of the other compounds were superconducting at temperatures above 4.2°K.

LITERATURE CITED

1. E. M. Savitskii and V. V. Baron, Izv. Akad. Nauk SSSR, Metallurgiya i Gornoe Delo, 5:3 (1963).
2. B. T. Matthias, T. H. Gebalde, and V. B. Compton, Rev. Mod. Phys., 35(1):1 (1963).
3. F. J. Morin and J. P. Maita, Phys. Rev., 129:1115 (1963).
4. A. M. Clogston and V. Jaccarino, Phys. Rev. 129:1357 (1963).
5. H. von Philipsborn, Mischsysteme von Verbindungen des Cr_3Si Typs und deren Polymorphie Erscheinungen, Zurich (1964).
6. B. W. Roberts, Progr. Cryogenics, 4:159 (1964).
7. Yu. V. Efimov, Izv. Akad. Nauk SSSR, Neorg. Mat., 2(4):605 (1966).
8. H. Holleck, H. H. Nowotny, and F. Benesovsky, Mh. Chemie, 94(2):473 (1963).
9. M. D. Banus, T. B. Reed, H. C. Gatos, M. C. Lavine, and J. A. Kafalas, J. Phys. Chem. Solids, 23:971 (1962).
10. W. R. Rossteutscher and K. Schubert, Z. Metallkunde, 55(10):617 (1964).
11. W. Piotrowski, Hutnik (Polska), 32(4):135 (1965).
12. B. T. Matthias, Progress in Low-Temperature Physics, Vol. II, Amsterdam (1957), p. 1.

SOME LAWS OF THE SUPERCONDUCTIVITY
OF INTERMETALLIC COMPOUNDS

G. V. Samsonov and E. N. Denbnovetskaya

An attempt is made at establishing a number of general laws governing variations in the superconducting characteristics of compounds of transition metals with other transition metals, with the so-called semimetals, and with certain nonmetals. A discussion is appended.

A large amount of experimental material has now been gathered together in relation to the superconducting transformation temperatures of various intermetallic compounds, compounds of metals with nonmetals [1], and alloys of intermetallic compounds [2] but as yet these have been inadequately discussed; perhaps the most important observation is the fact that superconductivity is exhibited most frequently by compounds containing transition metals [3]. In a previous investigation [4] we mentioned certain laws governing the superconductivity of refractory compounds of transition metals with nonmetals, both those of limiting composition and those lying within a range of homogeneity.

In this paper we shall attempt to present a number of superconductivity laws obeyed by intermetallic compounds (i.e., compounds between metals) and compounds of metals with the so-called semimetals and certain nonmetals by analyzing published data and considering the formation of stable electron configurations [5]. It is well known that in the modern theory of superconductivity the vanishing of the residual resistance, i.e., the transformation to superconductivity, is associated with the elimination of lattice defects which scatter the conduction electrons and also with the formation of an electron "quantum fluid" from coupled pairs of electrons. More specific conditions for the transformation into the superconducting state arising from the foregoing include the formation of stabler (i.e., statistically less easily disrupted) electron configurations in the atoms composing the crystal lattice by the components of alloys and compounds (or by the atoms of an element), (with the corresponding development of a special kind of "energy channel" with inert "walls"), together with an adequate carrier concentration [6]. The latter should not be excessively low, in order to avoid the development of semiconducting or dielectric properties, nor excessively high, in order to avoid strong interelectron interaction; other conditions being equal, the formation of pairs of conduction electrons is always favorable.

As shown earlier [5], the stable electron configurations of the d states include the d^0, d^5, and d^{10}, in which the degree of stability falls from d^0 to d^{10}, and the degree of stability of corresponding d configurations also falls with decreasing principal quantum number of the d electrons. In the case of sp elements it was shown that the sp^3 and s^2p^6 configurations had the greatest stability, the stability of corresponding configurations falling with increasing principal quantum number of the sp electrons. We may suppose that one of the foregoing conditions for the development of superconducting properties should be satisfied when the components of the

TABLE 1. Superconducting Transformation Temperatures for
Certain Compounds of Titanium, Zirconium, and Hafnium

Compound	T_K	Compound	T_K	Compound	T_K	Compound	T_K
TiOs	0.46	$ZrRu_2$	1.84	$ZrIr_2$	4.10	$ZrRe_2$	6.8
Ti_3Pt	0.58	Zr_3Sb_2	1.74	Zr_5Pb_3	4.60	$ZrRe_6$	7.40
TiCo	0.71	$HfOs_2$	2.69	$HfRe_2$	4.80	ZrV_2	8.8
Zr_3Al	0.73	ZrRh	2.7	Ti_3Ir	5.40	$TiNb_2$	9.3
Zr_3Pb	0.76	ZrPt	3.0	Ti_3Sb	5.80	$ZrTc_6$	9.7
Zr_3Au	0.92	$ZrOs_2$	3.0	$ZrRe_2$	6.0	$ZrNb_6$	10.8
TiRu	1.07	ZrB	2.8—3.2	Zr_2Co	6.30		

compounds form stable configurations, and the second condition should be satisfied when at the same time the carrier concentration is large enough or falls within certain limits.

Considering compounds of transition metals from this point of view (these have a high capacity for capturing electrons [7] and are far from forming stable d^5 electron configurations), we may assert that the values of T_K for these should, as a rule, be relatively low; this is in fact found to be so in the case of compounds of Ti and Zr and other transition metals with a small number of electrons in the d shell of the isolated atoms, except for those cases in which an extremely stable configuration of the partner occurs at the same time as the superconducting transformation (Table 1).

A large number of superconducting compounds are of course formed by niobium, particularly those of the Nb_3X type. It was shown in [8, 9] that of the five (d + s) valence electrons of niobium in the metallic crystal 3.7 to 3.8 electrons were localized and 1.2 to 1.3 existed in the collectivized state. We may suppose that the energetically most probable situation is the formation of stable d^{10} configurations by the niobium atoms, i.e., the combination of Nb atoms in threes (Nb_3) with the transfer of about one electron to the conduction band. This probably determines the tendency of niobium to form compounds chiefly of the form Nb_3X. According to existing data (all the T_K data in this paper are taken from the handbook [1]), T_K rises from 8.28 to 11.5°K on passing from Nb_3Ag to Nb_3Au. The high value of T_K may be explained by the existence of stable configurations of the d^{10} type, both for the groups of niobium atoms and for the silver and gold atoms, while the rise in T_K on passing from the silver to the gold compound may be explained by the reduction in the carrier concentration owing to the higher ionization potential of the gold atom, and also the greater probability of the transition of gold s electrons to the vacant f shell; at the same time, the energy stabilization of the d^{10} configurations of gold plays a decisive part owing to the larger quantum number of d electrons. In the series Nb_3Al, Nb_3Ga, Nb_3In T_K falls in the sequence 17.5, 14.5, 9.2°K; this correlates excellently with the fall in the energy stability of the sp^2 configurations of the atoms of the second component (formed as a result of the one-electron s → p transition $s^2p → sp^2$) and the simultaneous sharp rise in carrier concentration, leading to a strong interelectron interaction.

The sharp rise in T_K for Nb_3Sn is accordingly quite reasonable, since the condition for the formation of two stable d^{10} and sp^2 configurations is satisfied, and the number of conduction electrons facilitates the formation of pairs. Unfortunately, for this series of compounds the only other critical temperature known is that of Nb_3Ge (6.97°K), and at first glance this contradicts the foregoing tendency of T_K. It would appear likely that the extremely high energy stability of the sp^3 configuration causes a sharp statistical reduction in the concentration of conduction electrons on stabilization (this effect intensifies as the principal quantum number of the sp electrons becomes smaller), owing to the fact that these are used in the stabilization of the sp^3 configuration. It is evidently precisely because of this, that is, because of the strong individualization of the stable configurations and the small number of bond electrons, that compounds of the Nb_3C and Nb_3Si type are entirely absent.

For compounds of niobium with platinides (of the same Nb_3X composition), T_K rises in the sequence Nb_3Os, Nb_3Ir, Nb_3Rh, Nb_3Pt with values of 1.05, 1.7, 2.5, 9.20°K. In the case of Os $(5d^66s^2)$ the formation of stable d^5 configurations is the most probable occurrance, with the simultaneous appearance of a large number of conduction electrons, which leads to a low value of T_K as a result of the severe antibonding. In the case of Ir $(5d^76s^2)$ and Rh $(4d^85s^1)$ the statistical weight of the d^{10} configurations increases; although these are less stable than the d^5 in Os and Ir (their stability is further reduced in the case of Rh owing to the lower principal quantum number), their formation involves a sharp reduction in the carrier concentration and a diminution of antibonding. For platinum $(Nb_3Pt, 5d^96s^1)$ a high statistical weight of the d^5 configurations, a high principal quantum number, and the absence of excessively high carrier concentrations coincide favorably, and this yields the relatively high T_K of this compound.

At the same time, for the compounds OsNb, IrNb, RhNb, PtNb the critical temperatures are respectively 1.4, 7.9, 4.10, 2.4°K. The relatively low value of T_K for OsNb is due to the high carrier concentration, although this compound does form stable d^5-type configurations. In IrNb and RhNb there is a certain probability of the formation of d^{10} configurations as well as d^5. However, in RhNb the statistical weight of the d^{10} configurations is higher than in IrNb; in addition to this, their energy stability in RhNb is lower owing to the lower quantum number, so the T_K is lower for RhNb than for IrNb. For the compound NbPt, $T_K = 2.40$°K; this is due to the sharp reduction in the carrier concentration resulting from the rearrangement (at the expense of the carriers) of the electron configuration of one niobium atom to a configuration closely approaching d^5.

On the whole the foregoing considerations are supported on comparing the T_K of the Rh, Os, Ir, and Pt compounds; we clearly see (Table 2) a sharp rise in T_K on passing from Os to Ir, Rh, and Pt. We note that the relatively low values of T_K for the Os compounds are associated with the formation of stable d^5 configurations (which by itself should increase T_K) at the same time as severe antibonding of the electrons.

TABLE 2. Superconducting Transformation Temperature of the Compounds of Osmium, Iridium, Rhodium, and Platinum

Osmium $(5d^66s^2)$		Iridium $(5d^76s^2)$		Rhodium $(4d^85s^1)$		Platinum $(5d^96s^1)$	
Compound	T_K, °K	Compound	T_K, °K	Compound	T_K, °K	Compound	T_K, °K
OsTi	0.46	Ir_2Os_8	0.30	Rh_5P_4	1.22	PtSi	0.93
$OsNb_3$	1.05	$IrCr_3$	0.45	$RhTe_2$	1.51	PtTa	1.0
OsNb	1.40	IrTi	1.0	RhMo	1.97	$Pt_{0.5}W_{0.5}$	1.45
OsZr	1.5	IrSc	1.03	RhTa	2.0	$Pt_{0.3}Ta_{0.7}$	1.2—1.5
Os_3Th_7	1.51	Ir_3Th_7	1.52	RhBi	2.06	Pt_2J	1.57
Os_2Nb_3	1.78; 1.85	$IrNb_3$	1.7	Rh_5Ge_3	2.12	PtSb	2.10
OsTa	1.95	Ir_2Zr	4.10	Rh_3Th_7	2.15	PtNb	2.40
Os_2Nb	2.52	IrGe	4.70	$RhNb_3$	2.5	PtBi	1.21; 2.4
OsW_3	2.21—3.02	$IrTi_3$	5.40	$RhPb_2$	2.66	$PtPb_4$	2.80
$LuOs_2$	3.49	Ir_2Sr	5.7	RhZr	2.7	PtV_3	2.83
OsW	4.40	$CaIr_2$	4—6.15	$RhBi_3$	3.2	PtZr	3.0
OsMo	5.20	Ir_2Th	6.50	RhW	2.64—3.37	$PtNb_3$	9.20
		$IrMo_3$	6.8; 8.8	RhNb	4.10		
		IrNb	7.9	Rh_2Ba	6.0		
		$IrNb_3$	9.8	Rh_2Sr	6.2		
				Rh_2Ca	6.4		
				$Rh_{0.2}Zr_{0.8}$	9.0		

TABLE 3. Superconducting Transformation Temperature of Compounds of Elements Belonging to the Boron Subgroup [1]

Element	Compound	T_K, °K	Element	Compound	T_K, °K
B $2s^2 2p$	BRe_2	2.80	Tl $6s^2 6p$	TlAu	1.92
	BW_2	3.10		TlAg	2.67
	BTa_2	3.12		TlMg	2.75
	BZr	2.8—3.2		Tl_2Pb	3.75; 4.09
	BMo_2	4.74		$TlSb_2$	5.2
	BNb	8.25			
			In $5s^2 5p$	In_2Ag	2.3—2.46
Al $3s^2 3p$	$AlMo_3$	0.58		InTl	2.7
	$AlZr_3$	0.73		InHg	3.8
	Al_3Th	0.75		InPb	6.65
	Al_3Os	5.9		$InNb_3$	9.2
	$AlNb_3$	17.5; 18		$InLa_3$	10.4
Ga $4s^2 4p$	$GaMo_3$	0.76			
	GaV	7.6			
	Ga_2Mo	9.5			
	Ga_4Mo	9.8			
	GaV_4	10.1			
	$GaNb_3$	12.5; 14.5			
	GaV_3	16.8			

TABLE 4. Superconducting Transformation Temperatures for Compounds of Elements Belonging to the Nitrogen Subgroup [1]

Element	Compound	T_K, °K	Element	Compound	T_K, °K
N $2s^2 2p^3$	NTi	4.86; 5.60	Bi $6s^2 6p^3$	Bi_2Pd	1.70
	NMo_2	5		$BiAu_2$	1.74
	NNb	5.10; 15.6		Bi_3Ca	2.00
	NHf	6.2		BiRh	2.06
	N_3Nb_4	7.20		BiCu	2.20
	NV	7.50; 8.20		BiJ_3	2.25
	NZr	9.05; 10.7		BiNa	2.25
	NMo	12		BiPt	1.21; 1.24
P $3s^2 3p^3$	PW_3	2.26		Bi_3Rh	3.2
	PMo_3	5.31; 7		BiSn	3.48
	PPb	7.8		Bi_2K	3.58
As $4s^2 4p^3$	$AsPd_2$	1.7		BiK	3.60
	As_3Pd_5	1.9		BiPd	3.7; 4.25
	AsSn	4.10		BiSn	3.81
	AsPb	8.4		$BiPd_2$	4
Sb $5s^2 5p^3$	SbV_3	0.80		Bi_3Ni	4.06
	SbPd	1.25; 1.50		BiNi	4.25
	Sb_2Zr_3	1.74		Bi_2Pb	4.25
	Sb_4Mo_3	2.10		$BiIn_2$	5.6
	SbPt	2.10		Bi_3Sr	5.62
	Sb_2Sn_3	3.80		Bi_3Ba	5.69
	Sb_2Tl_7	5.20			
	Sb_2Tl	5.20			
	$SbTi_3$	5.80			
	SbPb	6.6			

On considering the T_K of compounds between elements of the boron subgroup and metals (or semimetals) (Table 3) we see that in general the T_K of Al compounds are lower than those of B compounds; this is legitimate since the stability of the sp^2 configuration is lower; compounds of Ga are characterized by the highest values of T_K, owing to the favorable combination of the relatively stable sp^2 configuration and an optimum carrier concentration; for In compounds T_K falls slightly, and this fall progresses further for compounds of Tl, owing to the decreasing stability of the sp^2 electron configurations and the simultaneous rise in carrier concentration.

Roughly the same picture occurs for compounds of elements belonging to the nitrogen subgroup (Table 4). According to [5] the transformation $s^2p^3 \rightarrow sp^3 + p$ may occur for nitrogen and members of its subgroup, i.e., ultimately a stable sp^3 configuration with a single p electron capable of passing to the partner may be formed. With the falling stability of the sp^3 configuration as the principal quantum number of the sp electrons increases, T_K becomes steadily smaller; exceptions occur in those cases in which there is a particularly high probability of a stable configuration forming in the atoms of the partner (for example, the nitrides of niobium, molybdenum, and vanadium). Here the p electron of the nitrogen makes up the defects of the d shell, tending to build it up into the corresponding stable state, thus ensuring the completion of the sp^3 configuration in the atoms of elements belonging to the nitrogen subgroup. In the case of carbon compounds, however, the tendency of the partner-atoms (particularly in the case of d elements) to be built up until they achieve stable electron configurations is realized at the cost of disrupting the stable configuration of the carbon atom, which leads to a fall in the T_K of the carbides as compared with the nitrides of the same metals; for example, the T_K of NbC is 10.3° and that of NbN 15.6°K; the T_K of MoC is 9.26°K and that of MoN 12°K. The same applies to other compounds of these subgroups; for example, the T_K of Mo_3Si is 1.30°K and that of MoP 5 to 7°K.

Transition metals with a high acceptor capacity (Ti, Zr, Hf, and partly V), as indicated earlier, cause disruption to the electron configurations of their partners, without, however, being very likely to form their own stable configurations, particularly d^5. Only in the case of nitrogen, where (as a result of the possibility of one p electron being transferred to an atom of the transition metal) the sp^3 configuration is maintained in its most complete form, are the transformation points of the corresponding compounds comparatively high. Thus, whereas for the silicides, borides, and carbides of Ti, Zr, and Hf the critical temperatures are extremely low [1], for TiN they lie between 4.86 and 5.60, for ZrN between 8.90 and 10.7, and for HfN at 6.2°K.

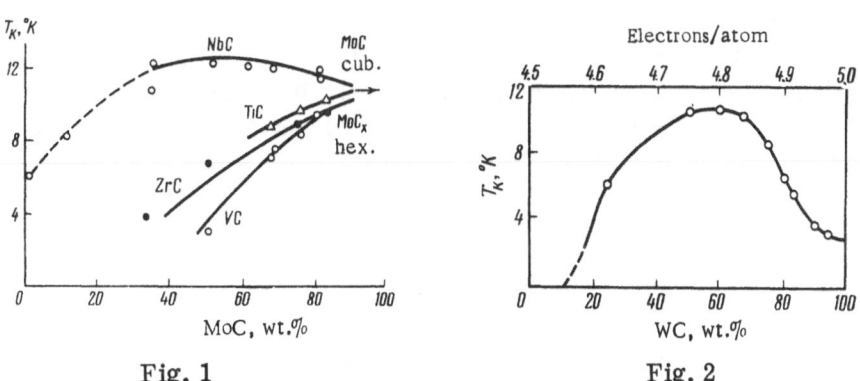

Fig. 1   Fig. 2

Fig. 1. Critical temperature of solid solutions of various carbides in molybdenum carbide of varying concentration.

Fig. 2. Critical temperature of the TaC—WC system as a function of composition [10].

In this connection the superconductivity data relating to solid solutions of various carbides in MoC [10] shown in Fig. 1 are particularly interesting. Taking the stable configuration sp^3 of the carbon atoms in these solid solutions as almost constant, the greatest T_K for the MoC–NbC system corresponds to the preferential formation of stable d^5 configurations for both the niobium $(4d^45s^1)$ and the molybdenum $(4d^55s^1)$ atoms, the maximum on the curve for this system in Fig. 1 evidently corresponding to the maximum probability of the formation of d^5 configurations by both atoms. On passing to the TiC–MoC system, owing to the high acceptor capacity of titanium, the electron configuration of the molybdenum is disrupted; however, the probability that the configuration of the titanium will accordingly be built up to d^5 is low, and as a result the total statistical weight of the d^5 configurations is in this case low compared with that of the MoC–NbC system, which leads to a fall in T_K ; the fall proceeds monotonically with increasing amounts of TiC in the solid solution, since these regularly lower the statistical weight of the d^5 configurations. This process goes further in the MoC–ZrC system, in view of the greater stability of the d configurations of zirconium as compared with those of titanium (rise in stability with increasing principal quantum number) and the corresponding greater disruption to the d^5 configurations of the molybdenum. The result is not so clear in the case of the MoC–VC; we should have expected a rise in the values of T_K as compared with the analogous system involving titanium carbide, owing to the greater probability of the formation of d^5 configurations by the vanadium atoms. It is possible that here, owing to the greater probability of a d^5 configuration being formed by the vanadium atom, not only the corresponding configuration of the molybdenum atom but also the sp^3 configuration of the carbon atom is disrupted.

It is clear from the foregoing that the concept (partly used by Matthias and other research workers) of the number of electrons per atom as a characteristic giving some idea of the possibility of a superconducting transformation taking place and of the magnitude of T_K is theoretically devoid of meaning. Thus, on the basis of the same paper [10], Fig. 1 shows the value of T_K for the TaC–WC system as a function of the concentration of valence electrons per atom. It is clear that the maximum value of T_K corresponds to the maximum statistical weight of the d^5 configurations formed by the tantalum and tungsten atoms; owing to the disruption of the d electron configuration of tungsten in order to form stable configurations in the tantalum atoms, the maximum temperature of the transformation is displaced in the direction of larger concentrations of tungsten carbide. The fall in T_K from the maximum in the direction of increasing TaC content is due to the reduced probability of the formation of d^5 configurations by the tantalum atoms, while the fall from the maximum in the direction of WC is associated with the reduction in carrier concentration.

It was shown in [11] that in the V_3Si–Mo_3Si system (V_3Si has a high critical temperature equal to 17°K) there is a sharp fall in T_K on adding small amounts of molybdenum (up to 5%), then a smoother fall in T_K on increasing the Mo_3Si content. It is clear that the reduction in critical temperature is due to the sharp rise in carrier concentration in the case of Mo_3Si, in which molybdenum groups, having an excess of carriers, combine with Si atoms, the configurations of which are insufficiently stable, and in turn break up, with a considerable increase in carrier concentration.

As already indicated, together with the requirement of maximum stability in the electron configurations of the crystal-lattice components, an important condition for passage into the superconducting state is a certain concentration of conduction electrons, such that exceeding this concentration would lead to severe antibonding, and failing to reach it would lead first to semiconducting and then to insulating properties. Since certain specific boundaries are involved in this argument, it is evident that substances lying at these boundaries (or situated in certain fields effectively constituting boundaries) may simultaneously possess semiconducting properties and be capable of passing into a state of superconductivity. Thus for compounds of the $A^{III}B^V$ type (A = B, Al, Ga, In, Ti; B = N, P, As, Sb, Bi) we may consider the compounds of the A^{III} elements with nitrogen. In all cases the A^{III} element, in the isolated state, has an atom of configuration s^2p, transforming by way of an $s \rightarrow p$ transformation into sp^2, while the nitrogen, for reasons indicated earlier, passes into the sp^3 configuration with a p electron capable of bing transferred to the A^{III} atom, with formation of a stable sp^3 configuration. In other words, two stable sp^3 configurations are formed, and the energy stability of these falls with increasing principal quantum number of the sp electrons. In view of this the first member of the series, BN, constitutes a compound with a wide forbidden band (about 5 eV), characteristic of dielectrics. In view of the individualization of the sp^3 configurations of the components, in addition to a wide forbidden band BN has a number of specific properties, for example, a tendency toward dissociation on evaporating. For the next member of the series, AlN, the width of the forbidden band is smaller (3.8 eV), owing to the fall in the energy stability of the sp^3 configuration of Al as compared with that of B and the consequent intensification of electron exchange between the components, while the tendency toward dissociation into the elements on evaporation is maintained, although at relatively higher temperatures (there is a slight fall in the individualization of the sp^3 configurations). On passing to gallium nitride, a typical semiconductor ($\Delta E = 3.25$ eV), there is hardly any dissociation on evaporating [12]; at the same time this compound is a superconductor with a relatively high T_K [13]. In the same way indium nitride while remaining a semiconductor ($\Delta E = 2.4$ eV), undergoes a transformation to superconductivity at 3.23°K, i.e., both nitrides (GaN and InN) are in the transitional region of carrier concentration in which semiconducting and superconducting properties may coexist.

Thus by using the concepts of the theory of stable electron configurations we may quite convincingly explain (at least qualitatively) the laws governing the variations in T_K observed for various types of chemical compounds, and this in turn should enable us to find reliable ways of alloying technically-important superconductors and finding new compounds with superconducting properties.

LITERATURE CITED

1. B. W. Roberts, Superconductive Materials and Some of Their Properties, General Electric Res. Lab., New York (1963).
2. I. I. Kornilov, The Chemistry of Metallides, Consultants Bureau, New York (1966).
3. E. M. Savitskii and V. V. Baron, Izv. Akad. Nauk SSSR, Metallurgiya i Gornoe Delo, No. 5, p. 1 (1964).
4. G. V. Samsonov, in: "Metallography and Metallophysics of Superconductors" [In Russian], Izd. "Nauka" (1965), p. 65.
5. G. V. Samsonov, Ukr. Khim. Zh., 31(12):1233 (1965).
6. I. M. Chapnik, Dokl. Akad. Nauk SSSR, 141:70 (1961).
7. G. V. Samsonov, Dokl. Akad. Nauk SSSR, 93:689 (1953).
8. M. I. Korsunskii and Ya. E. Genkin, Dokl. Akad. Nauk SSSR, 142:1276 (1962).
9. M. I. Korsunskii and Ya. E. Genkin, Dokl. Akad. Nauk SSSR, Ser. Fiz., 28:832 (1964).
10. B. T. Matthias, E. A. Wood, E. Corenzwit, and V. B. Bala, J. Phys. Chem. Solids, 1:188 (1956).
11. N. E. Alekseevskii, E. M. Savitskii, V. V. Baron, and Yu. V. Efimov, Dokl. Akad. Nauk SSSR, 154:82 (1962).
12. S. P. Gordienko, G. V. Samsonov, and V. V. Fesenko, Zh. Fiz. Khim., 38:2974 (1964).
13. N. E. Alekseevskii, G. V. Samsonov, and O. I. Shulishova, Zh. Éksp. Teor. Fiz., 44:1413 (1963).

DISCUSSION ON
THE PAPER OF G. V. SAMSONOV AND
E. N. DENBNOVETSKAYA

V. V. Shmidt (Baikov Institute of Metallurgy). I should like to say a few words on Samsonov and Denbnovetskaya's paper. I am dissatisfied with this contribution for several reasons. Firstly there are no positive contents. This is most important. I was unable to capture any specific assertions. Secondly (and these are subsidiary points), the terminology employed is quite impermissible. Some "channels," some "walls." It is essential to stipulate that a certain physical terminology exists and this must be adhered to; if everyone invents new concepts, we shall deplete the whole language. Finally (third point), the categorical assertion that Matthias was wrong and that his rules have neither theoretical nor experimental bases is untrue.

A treatise on the theory of superconductivity was published in 1960; this gave a basis for the Matthias rules (in an article by Pines) so that there is in fact a theoretical basis for these. It is perhaps true that there may be deviations from these rules, since the theory was based on quite simple considerations, whereas the affair is really more complex; at any rate, it is quite wrong to reject these matters from all consideration.

I feel that the standard of such contributions should be improved.

Ya. S. Umanskii (Moscow Institute of Steels and Alloys). I should like to add something to what the previous speaker has said and make some comments on the actual concept of these stable electron configurations.

It must be emphasized that a stable configuration may be ascribed to any arbitrary phases whatsoever by using the methods employed by the authors. The authors propose accepting a stable configuration of ten electrons. In the same way one might consider that any element with a ten configuration was a superconductor.

When one says that three particular atoms form a stable configuration in their d shells, then clearly these three atoms should lie at specific distances from each other. By working with these distances one might prove anything one likes.

When one considers three niobium atoms, what does this really mean? Is this an odd sort of Nb_3 molecule? Where does this molecule lie in the lattice? The structure of Nb_3X is well known. However, there are no groupings of three transition-metal atoms with reduced distances. Of course all this is pure speculation, independently of what it is applied to, be it a question of melting points, greater hardness, or superconductivity. In the authors' opinion, high melting points are also due to the same groupings. Take a number of atoms, say that they have d shells, and, behold, everything follows. You say ten; I might suggest six electrons. And everything happens exactly as before. It might be five or four, or any combination. This arbitrary approach has neither a theoretical nor an experimental basis.

V. PHASE DIAGRAMS OF SUPERCONDUCTING ALLOYS

PHASE DIAGRAM OF THE Nb—Sn SYSTEM

V. G. Kuznetsova, V. A. Kovaleva, and A. V. Beznosikova

The phase diagram of the Nb-Sn system is studied by metallographic analysis, x-ray diffraction, x-ray spectral microanalysis, and also thermal and chemical analysis. The alloys are studied in the cast and quenched states and also after prolonged annealing (up to 2000 h). There intermetallic compounds occur in the system: Nb_3Sn (β phase, about 75 at.% Nb), Nb_3Sn_2 (γ phase, about 58 at.% Nb), and $NbSn_2$ (δ phase, about 33.5 at.% Nb). The compounds Nb_3Sn_2 and $NbSn_2$ are formed by a peritectic reaction at $910 \pm 10°$ and $840 \pm 10°C$ respectively. The compounds Nb_3Sn and $NbSn_2$ are stable down to room temperatures; the range of stability of Nb_3Sn_2 lies between 820 and 910°C. The crystal structure of the δ phase is studied for the case of single crystals. The compound has an orthorhombic structure with parameters a = 5.65 Å, b = 9.85 Å, and c = 19.2 Å. The structure of the γ phase has not been established. There is a range of homogeneity in the compound Nb_3Sn; at 950°C the solubility of tin in Nb_3Sn equals 0.6 at.% and that of niobium 1.3 at.%; at 1500°C these values rise to 1.02 at.% Sn and 1.0 at.% Nb. The solubility of tin in solid niobium at 1000 to 1500°C is no greater than 2.5 at.% Sn.

The first information regarding the properties of Nb-Sn alloys was published in 1954 by Matthias [1], who found an intermetallic compound Nb_3Sn having a structure of the β-W (A15) type and containing 29.87 wt.% Sn. It was considered that this compound was formed by a peritectic reaction between about 1200 and 1500°C. Later, in 1955, Geller [2] made a more accurate determination of the lattice constant of Nb_3Sn: a = 5.289 ± 0.002 Å. Both in the first and the second paper it was found that Nb_3Sn had the highest known superconducting transformation temperature (18.05°K). The superconducting properties of this compound were studied in [3, 4].

The first phase diagram of the Nb—Sn system appeared in 1959 and seven versions have now been published. These diagrams differ in the number of phases present, their composition and ranges of stability, the solubility of tin in niobium, the range of homogeneity of Nb_3Sn, and so on. The appearance of this large number of versions of the phase diagram and the differences between them bear witness to the complexity involved in studying Nb—Sn alloys; this arises from the great difference in the melting points of tin and niobium, the high vapor tension of tin above 1000°C, and the slowness of the diffusion processes at lower temperatures.

The Nb—Sn phase diagram was first developed by Soviet scientists on the basis of cast alloys [5], using microstructural and thermal analysis and x-ray diffraction. It was found that only one compound, Nb_3Sn, was stable at low temperatures in this system.

Another phase diagram differing fundamentally from the former was published in 1962 [6]. In this paper a metallographical and x-ray spectral analysis of metalloceramic alloys was presented and four intermediate intermetallic compounds were found: Nb_4Sn (t_m = 2040°C), Nb_3Sn (t_m = 730°C), Nb_2Sn (t_m = 690°C) and Nb_2Sn_3 (t_m = 893°C). All the compounds were stable down to room temperature.

Later Nb–Sn compounds were again studied [7]; on the basis of an x-ray structural analysis of the diffusion layers formed between the tin and niobium the existence of two intermediate intermetallic compounds was established, one of these being Nb_3Sn, while the composition of the other was not accurately determined.

In the following two years four more papers appeared [8-11]; the general indication from these was that the system held three intermediate phases. However, apart from this common characteristic the papers in question contained great differences in the temperature ranges corresponding to the existence of the various phases and the compositions of the latter. Thus in [8, 9, 11] Nb_3Sn decomposed at low temperatures; however, the decomposition temperature varied from 775°C [9] to 875°C [8]. The assertion that Nb_3Sn was subject to decomposition was based on the results of the comparatively brief, low-temperature sintering of metalloceramic mixtures of tin and niobium powders taken in the ratio of Nb_3Sn, after which only the low-temperature tin-rich δ phase was found in the alloys.

This phase is regarded as being stable down to room temperatures in all the papers. According to [8-10] the composition of this phase corresponds to the compound Nb_2Sn_3; according to [11] it may be $NbSn_2$.

There is also some disagreement as regards the range of stability of the γ phase. The composition of this corresponds to the compound Nb_3Sn_2, according to [8-11].

The range of solubility of tin in solid niobium was studied in all the papers except [8, 11]; however, the limiting solubility and its temperature dependence were treated differently.

In all the papers except [5, 7], metalloceramic alloys (obtained by sintering powder mixtures) and diffusion layers were studied. No cast alloys were examined in these cases.

However, owing to the slowness of the diffusion processes, particularly those occurring between niobium and the intermetallic phases arising at the onset of sintering, metalloceramic alloys usually have a nonequilibrium structure, very hard to solve by metallographic methods, owing to the severe porosity of the samples and the fragility of the phases.

In view of this it is quite insufficient to study the Nb–Sn system on the basis of metalloceramic alloys only. In the present investigation we therefore supplemented a wide study of such alloys by an examination of the structure of cast alloys after various forms of heat treatment and also by an examination of the structure of diffusion layers.

We used metallographic, x-ray structural, thermal, and chemical methods as well as the x-ray microanalysis of microsections. The alloys were melted in an arc furnace in an argon or helium atmosphere. In preparing the materials for the alloys allowance was made for the severe evaporation of the tin on melting. The structure of alloys was studied in the cast, annealed, and quenched states. Since the processes leading to the equilibrium state of the Nb–Sn alloys only occurred very slowly, the alloys were heated over a long period (up to 2000 h) in the course of heat treatment. The samples were annealed in quartz ampoules (up to 900 or 1000°C) or in hermetic niobium containers (above 1000°C) filled with argon.

The phase compositions of alloys from all parts of the Nb–Sn phase diagram were determined as a result of this work.

Alloys between 3 and 25.7 at.% in the cast state (Fig. 1a) have a two-phased structure consisting of a solid solution on niobium base (α phase) and the compound Nb_3Sn (β phase). Annealing alloys from this region at temperatures between 550 and 1500°C for up to 2000 h produced no change in phase composition. The β phase remained stable down to room temperature; however, the Nb-base α solid solution decomposed (Fig. 1b). The thermal analysis of alloys in this region confirmed the absence of any phase transformations.

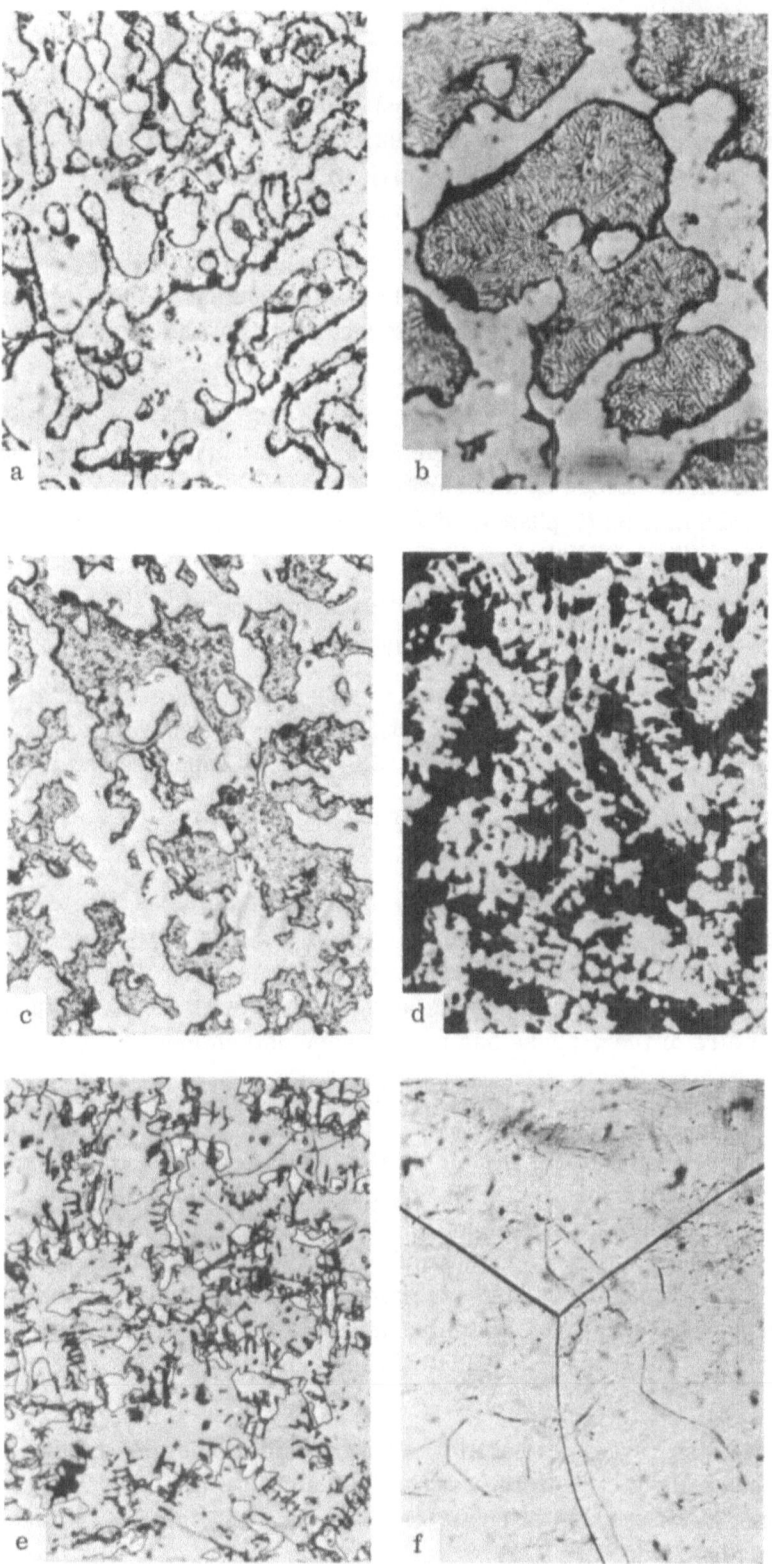

Fig. 1. Microstructure of Nb–Sn alloys. a) 18.1% Sn, cast
(× 340); b) the same after annealing for 120 h at 1000° (× 1000);
c) 59% Sn, cast (× 340); d) the same after annealing for 500 h at
700° (× 340); e) the same after annealing for 36 h at 900° and
water quenching; f) 3% Sn, cast, α phase (× 140).

All the alloys with over 25 at.% Sn had a structure comprising Nb_3Sn and tin after casting (Fig. 1c). Annealing such alloys at low temperatures (up to 910°C) leads to the appearance of new intermetallic compounds (δ and γ phases). The appearance of the δ phase was observed after the prolonged annealing of cast alloys in this range at 700 and 800°C for 500 h or 550°C for 2000 h (Fig. 1d). After etching in an acid mixture (HNO_3:HCl 1:1) the δ phase adopted a dark gray color. The microhardness of this phase was 600 kg/mm^2.

The appearance of the γ phase was observed on annealing cast alloys (25 to 67 at.% Sn) at 820 to 910°C (Fig. 1e) with subsequent quenching. In contrast to the δ phase, the γ phase acquires a light-gray color on etching. The microhardness of the γ phase is 100 kg/mm^2 higher than that of the δ phase.

X-ray spectral analysis of the microsections showed that the tin content of the two phases was about 40% for the γ phase and 67% for the δ phase. According to chemical analysis, these phases (separated from the alloys by dissolution of the excess tin in hydrochloric acid) contain 42% Sn (γ phase) and 66.5% Sn (δ phase). It follows that the γ phase corresponds to the compound Nb_3Sn_2 and the δ phase to $NbSn_2$.

On heating the samples above 910°C the structure of the alloys is analogous to that of the cast alloys of this region and comprises the β phase and tin.

Thermal analysis of alloys containing 29.5 and 44.9 at.% Sn annealed at 700°C for 500 h revealed the presence of three rests on the heating and cooling curves, corresponding to the temperatures 820, 840, and 910°C (the rests at 820 and 840°C appeared as a double peak).

It is well known that Nb_3Sn has a cubic structure of the Cr_3Si type with a lattice constant of a = 5.289 Å.

In order to determine the structure of the δ phase we obtained single crystals of this compound. The compound had an orthorhombic structure. The lattice constants found from x-ray diffraction pictures based on rotation around three mutually perpendicular axes were a = 5.65; b = 9.85; c = 19.2 Å. These results agree closely with the latest in the literature [12, 10]. The structure of the γ phase has not yet been determined.

The lattice constant of Nb_3Sn varies from 5.282 to 5.292 Å depending on the mode of heat treatment applied to the alloy; this indicates a range of homogeneity for this compound. In order to determine the solubility of tin and niobium in Nb_3Sn the latter was separated from alloys containing excess tin, which had been subjected to heat treatment at various temperatures (950, 1000, 1500°), by dissolving the excess tin in HCl (until the tin lines disappeared from the diffraction picture). The residual powder, constituting the compound Nb_3Sn, was subjected to x-ray structural and chemical analysis. On the basis of these experiments it was established that the range of homogeneity of Nb_3Sn expanded on raising the temperature. At 950° the solubility of tin in Nb_3Sn is 0.6 at.% and that of niobium 1.3 at.%; at 1500° these values increase to 1.02 and 1.9 at.% respectively.

We also studied the range of solubility of tin in solid niobium. For this purpose we prepared a series of cast alloys containing between 0.08 and 12.8 at.% Sn. The cast alloys were subjected to a homogenizing anneal at 1800° for 20 h. The structure of the cast alloys up to 3% Sn remained single-phase (Fig. 1f).

As a result of annealing the alloys at 1000 and 1500° it was established that at these temperatures the solubility of tin in niobium was no greater than 2.5 at.%, since the alloy containing this amount of tin showed the appearance of a second phase (the β phase) after annealing (Fig. 2a).

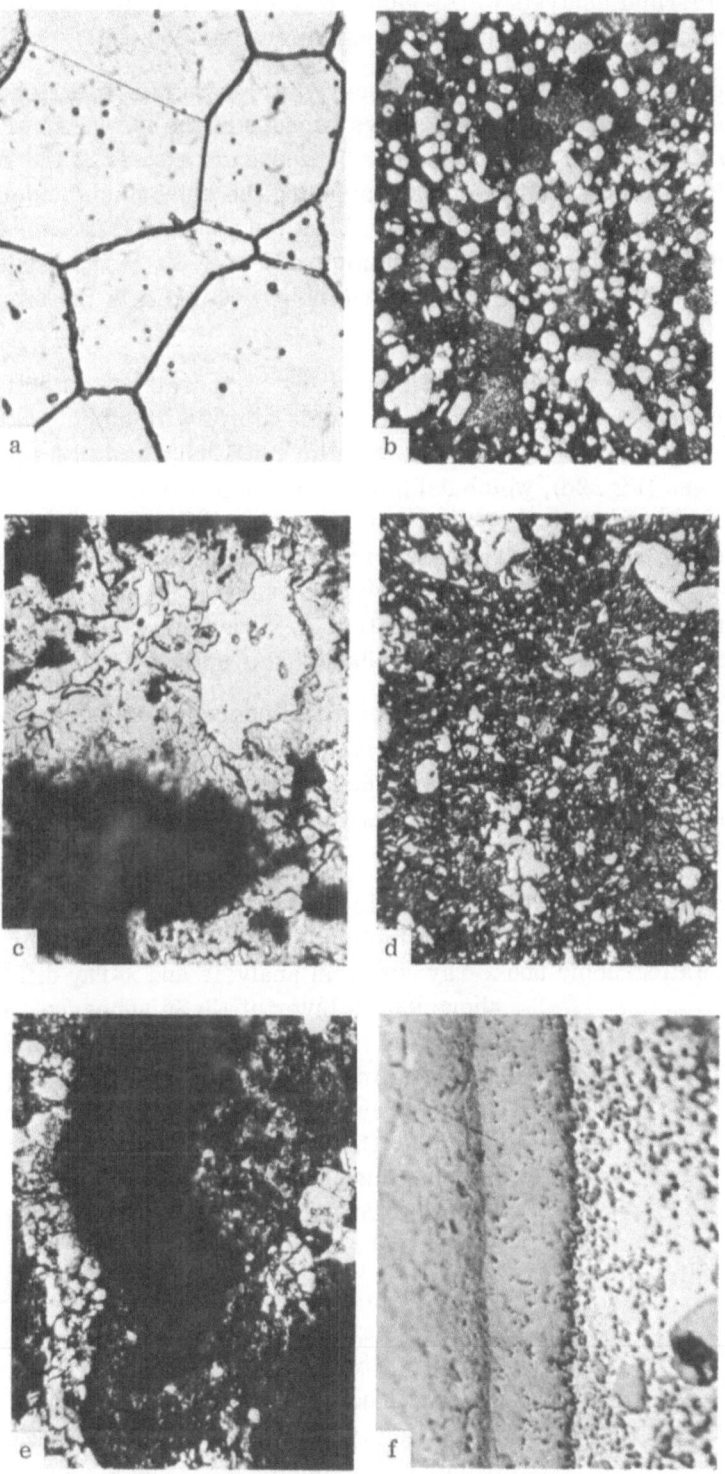

Fig. 2. Microstructure of Nb–Sn alloys. a) Alloys with
2.5% Sn after annealing for 120 h at 1000° (× 140); b) alloy with 70% Sn, cast (× 200); c) metalloceramic alloy
with 50% Sn after annealing for 2000 h at 850° (× 520); d)
metalloceramic alloy with 25% Sn after annealing for 120
h at 1000° (×140); e) the same after annealing for 700 h
at 650° (× 230); f) structure of a diffusion layer annealed
for 500 h at 650° (× 140).

By applying thermal analysis to tin-rich alloys (over 70% Sn, Fig. 2b) the temperature of the peritectic reaction $NbSn_2 + L$ (Sn), was established as $234 \pm 1°$.

We also studied the structure of metalloceramic Nb-Sn alloys, which in a number of cases differed from the structure of cast alloys because of the slow rate of the reactions taking place in the sintering of Nb-Sn powder mixtures. However, a study of the structure of such alloys sintered at 650 to 1500°C helped in understanding the mechanism underlying the formation of the phases and the interaction of the latter. Thus the γ phase (Fig. 2c), formed as a result of annealing at 850° for 2000 h in an alloy containing 50 at.% Sn, is the product of a reaction between the β phase (white coloring) and tin originally occurring at the site of the existing pores.

In order to verify the instability of Nb_3Sn at low temperatures established in [7, 8, 10] we carried out some experiments on metalloceramic alloys with a composition corresponding to the compound Nb_3Sn. Annealing one of these alloys at 1000° for 120 h leads to the formation of the compound Nb_3Sn (Fig. 2d), which exhibits no decomposition into other phases on subsequent annealing at 650° for 2000 h. If, however, this alloy is first held at 650° for 700 h, the compound Nb_3Sn is never formed. The structure of the alloy in this case consists of the α solid solution of niobium and the low-temperature δ phase (Fig. 2e). As a result of further annealing at the same temperature (650°) for 1500 h, x-ray analysis of the alloy reveals traces of Nb_3Sn in addition to the earlier-formed δ phase and α solid solution.

Thus in the low-temperature sintering of metalloceramic alloys the equilibrium state is reached in two stages: First by the formation of the compound $NbSn_2$, and then by the conversion of this into Nb_3Sn. Experiments showed that the second stage of the reaction took place at an extremely slow rate, and x-ray analysis only revealed the development of Nb_3Sn after prolonged annealing (up to 2000 h).

In parallel with our study of the structure of cast and metalloceramic alloys, diffusion layers formed on a niobium wire placed in molten tin and vacuum-annealed at 650 to 1000°C were studied by metallographic and x-ray spectral analysis and x-ray diffraction. On heating above 950° a layer of Nb_3Sn appeared. Below this temperature and right down to room temperature the δ phase was observed in the diffusion layer (Fig. 2f). The γ phase only appeared in the diffusion layers on quenching the latter from 850°C. The amount of tin in the layers (according to the chemical and x-ray spectral analysis of the microsections) corresponded to the compositions of the compounds Nb_3Sn_2 and $NbSn_2$.

Thus the phase diagram of the Nb-Sn system (Fig. 3) is characterized by the existence of three intermetallic compounds: Nb_3Sn (β phase), Nb_3Sn_2 (γ phase), and $NbSn_2$ (δ phase). The crystal structures of the compounds Nb_3Sn and $NbSn_2$, known from earlier determinations, were confirmed by our own experiments. The structure of Nb_3Sn_2 has still not been solved.

On analysing the results obtained by the prolonged annealing of cast and metalloceramic alloys and also thermal-analysis data, we find that Nb_3Sn is stable down to room temperature and has a range of homogeneity extending from 23.1 to 26.02 at.% Sn at 1500°C. The range of homogeneity contracts with falling temperature. The temperature range

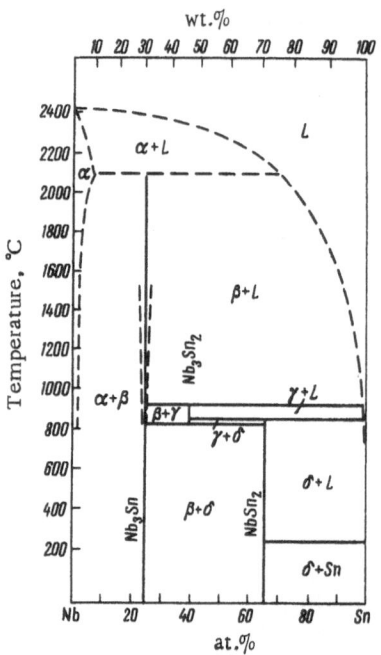

Fig. 3. Phase diagram of the Nb-Sn system.

corresponding to the stability of Nb_3Sn_2, 820 to 910°C, is deduced by studying the corresponding alloys after heat treatment (prolonged annealing with subsequent quenching from this range of temperatures), and also from the existence of rests on the thermal-analysis curves. However, the structure of the alloys annealed at temperatures below 820°C varies from one experiment to another and gives no unequivocal solution to the question as to whether this compound develops instability before reaching room temperature. This question will be studied again later.

LITERATURE CITED

1. B. T. Matthias, T. H. Geballe, S. Geller, and C. Corenzwit, Phys. Rev., 95:1435 (1954).
2. S. Geller, B. T. Matthias, and R. Goldstein, J. Am. Chem. Soc., 77(15):21 (1955).
3. R. Enstrom, T. Courtney, G. Pearsall, and U. Wulff, in: Metallurgy of Advanced Electronic Materials (G. E. Brock, ed.), Gordon and Breach, New York (1963).
4. T. B. Reed, H. C. Gatos, W. J. Lafleur, and J. T. Roddy, ibid.
5. M. N. Agafonova, V. V. Baron, and E. M. Savitskii, Izv. Akad. Nauk SSSR, Otd. Tekh. Nauk, Metallurgiya i Toplivo, No. 5, p. 138 (1959).
6. G. L. Wyman, I. R. Cuthill, G. A. Moorr, I. I. Park, and H. Yakowitz, J. Research Natl. Bur. Stands., 66A:351 (1962).
7. V. S. Kogan, A. I. Krivko, B. T. Lazarev, L. S. Lazareva, A. A. Matsakova, and O. N. Ovcharenko, Fiz. Met. Metallov., 15(1):143 (1963).
8. T. B. Reed, H. C. Gatos, W. J. Lafleur, and J. T. Roddy, Superconductors, Interscience, New York (1962).
9. R. E. Enstrom, G. W. Pearsall, and J. Wulff, J. Metals, 16:97 (abstract) (1964).
10. T. G. Ellis and H. A. Wilhelm, J. Less-Comm. Metals, 7(1):67 (1964).
11. H. J. Levinstein and E. Buehler, Trans. Metallurgical Soc., 23(6):1314a, 1321 (1964).
12. D. J. Ooijen, J. H. N. Van Vucht, and W. F. Van Druyvesten, Phys. Lett., 3(3):128, 129 (1962).

PHASE DIAGRAM OF THE Nb—Sn SYSTEM

V. N. Svechnikov, V. M. Pan, and Yu. I. Beletskii

The phase diagram of the Nb–Sn system is presented. In the equilibrium state three intermetallic compounds are formed: Nb_3Sn, Nb_6Sn_5, and $NbSn_2$. The compound Nb_6Sn_5 is stable over a narrow temperature range (815 to 920°) and $NbSn_2$ is stable from 860° to room temperature. It is suggested that phase separation takes place in the liquid state and that there is a monotectic equilibrium at 960°.

In an earlier paper [1] devoted to the Nb–Sn phase diagram we studied the range of existence of the β phase, which had a crystal lattice of the β-W (A15) type and the stoichiometric composition of Nb_3Sn. We found that the β-Nb_3Sn phase was stable right down to room temperature and had a wide range of homogeneity (about 18 at.%). The solubility of niobium in Nb_3Sn falls monotonically with increasing temperature, while on raising the temperature the solubility of tin first rises, reaching nearly 32 at.% at 900°C, and then starts falling. Since the critical superconducting-transformation temperature of the β-Nb_3Sn phase rises linearly with increasing tin concentration, the maximum T_K may be attained for a β-Nb_3Sn phase containing 32 at.% Sn, for example, in a diffusion layer obtained after annealing at 900°C.*

By studying the literature relating to the Nb–Sn system† we find that there are very serious contradictions in determining the form of the phase diagram of the tin-rich part of the Nb–Sn system. Nor are we satisfied by the results of some new investigations [2, 3] not mentioned in our review of the literature. The conclusions of [2] in essence repeat the conclusions of Ellis and Wilhelm [4], although the compounds in question are named differently: Nb_6Sn_5 and $NbSn_2$ instead of Nb_3Sn_2 and Nb_2Sn_3. Anantharaman [3] proposed the formula Nb_9Sn_7 for a compound which (according to his data) had the composition 43 to 44 at.% Sn and a tetragonal lattice. However, analysis of the literature shows that this compound would be better described as Nb_6Sn_5 (45.45 at.% Sn) and that its unit cell is orthorhombic, of the β-Ti_6Sn_5 type, with lattice constants a = 5.6549, b = 9.2057 and c = 16.814 Å (see [3] and the literature review mentioned). As regards the compound containing a higher proportion of tin, this would appear to be generally accepted as $NbSn_2$, having an orthorhombic cell of the $CuMg_2$ type with lattice constants a = 5.6450, b = 9.8576 and c = 19.121 Å (see the same literature).

The temperatures ranges of stability of these compounds and hence the general form of the phase-equilibrium diagram of the Nb–Sn system still remain contentious and have never been reliably established.

---

* It will later be shown that this temperature is about 960°C.

† See the article "Reasons for Changes in the Superconducting Properties of Nb_3Sn," this volume, p. 117.

In the present investigation we set ourselves the problem of plotting the phase diagram of the tin-rich part (over 40 at.% Sn) of the Nb—Sn system and thus completing our study of this system started earlier.

The original materials for the preparation of the alloys were niobium in the form of short rods (99.6% Nb) and tin bars (99.9995% Sn).

Alloys containing 40% of tin or over (See Table 1) were prepared from a mixture of Nb_3Sn powder (melted in an arc furnace and then crushed in a mortar) and pure tin shavings. This mixture was melted once in an arc furnace, so producing compact bars in which Nb_3Sn particles were fairly evenly distributed in the tin. Then for annealing the resultant bars were pressed into niobium cylinders with screw tops. The niobium cylinders were placed in a double quartz ampoule (evacuated and then filled with argon). Annealing was carried out at 750°C for 700 h; during the annealing the ampoule was continuously rotated at a rate of 2 rpm in order to prevent liquation with respect to specific gravity.

The principal methods of investigation were differential thermal analysis (the construction of the apparatus was described in [5]) and x-ray structural analysis. Microstructural analysis was employed as a supplementary technique. The x-ray diffraction pictures were taken with rotating powder samples in a cylindrical camera of diameter 57.3 mm, using unfiltered chromium radiation.

The results are shown in Fig. 1 in the form of a complete diagram of the phase equilibria of the Nb—Sn system. Here the horizontal corresponding to the peritectic equilibrium $\alpha + L \rightleftharpoons \beta$, is drawn at 2130°C on the basis of the results given in [4, 6], the line of the liquidus from the results of [2, 6], and the line representing the solubility of tin in niobium from the results of [2, 4, 7, 8]. The range of homogeneity of the β phase based on the compound Nb_3Sn with the β-W structure is taken from the results of our own earlier work.

Differential thermal analysis of the alloys containing over 40 at.% Sn annealed at 750°C for 700 h showed sharp thermal effects at 815, 860, 920, and 960°C on heating (see examples of thermograms in Fig. 2); whereas the thermogram of the alloy containing 42.9% Sn (Fig. 2a) shows two effects on heating, at 815 and 915°C, and the thermograms of the alloy containing 65.0% Sn (Fig. 2d) also show two effects, at 860 and 955°C, the thermogram of, for example, an alloy containing 53.2% Sn (Fig. 2b) clearly shows four effects, at 820, 870, 920, and 970°C. The thermal effect of the melting of tin only appears on the heating thermograms for alloys with

TABLE 1. Composition of the Alloys Studied*

Number of alloy	By charge, % Sn		By analysis, % Sn	
	at.	wt.	at.	wt.
13	40	46.0	42.9	49.0
14	45	51.11	52.4	58.4
15	50	56.09	53.2	59.2
16	60	65.71	56.0	61.9
17	66.7	71.87	54.4	60.4
18	70	74.88	60.0	65.7
19	70	74.88	63.6	69.1
20	75	79.31	65.0	70.3
21	80	83.63	71.9	76.6
22	85	87.87	79.3	83.0
23	90	92.0	77.6	81.6

*There was a considerable scatter among parallel samples in the chemical analysis of some alloys. In these cases the average composition was taken from five parallel samples.

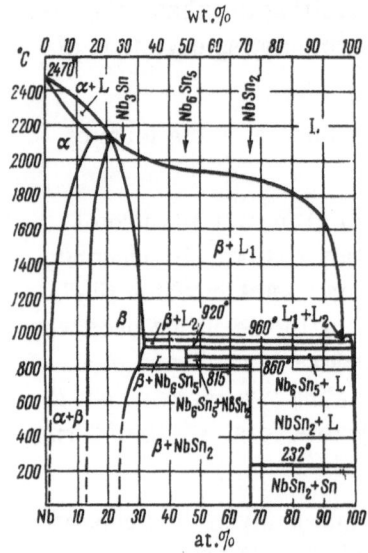

Fig. 1. Phase diagram of the
Nb–Sn system.

71.9, 77.6, and 79.3% Sn (see, for example, Fig. 2e); in alloys with a lower tin content, including that with 65.0% Sn, there is no such effect (Fig. 2d). On either cooling or reheating there are two thermal effects in all the alloys: on cooling, a thermal effect at 790 to 860°C (the position mainly depends, as it would appear, on the cooling rate) and the effect of the crystallization of tin at 180 to 200°C; on second heating, the effect of the tin melting at 230°C and an effect at about 860°C.

In order to elucidate the results of the thermal analysis we made an x-ray structural analysis of these alloys. We took x-ray diffraction pictures after annealing at 750°C for 700 h. In all the alloys x-ray diffraction revealed the $NbSn_2$ phase. Free tin only appeared in alloys with 71.9, 77.6, and 79.3% Sn. The β-Nb_3Sn phase appeared in alloys with a tin content of up to 60.0%; for higher concentrations of tin (up to 65%) the β phase was evidently present in such quantities that x-ray analysis was "insensitive" to it. These results show that at 750°C only two intermediate phases exist in Nb–Sn alloys in the equilibrium state; one of these

is β-Nb_3Sn, the composition of the second intermediate compound corresponding to the formula $NbSn_2$. This agrees with the x-ray and thermal analyses: Tin is present in the 65.0%-Sn alloy but absent from the alloy with 71.9% Sn when these alloys are brought into a state of equilibrium.

In order to discover what type of nonvariant equilibria correspond to the horizontals at 815, 860, 920, and 960° (± 10°C) plotted from the results of thermal analysis, alloys Nos. 13, 14, 15, 16, 18, 21 were quenched from 790, 820, 860, 880, 930, and 1000°C. The x-ray structural analyses of alloys Nos. 14 and 16 quenched from 790°C only showed a slight increase in the comparative intensity of the β-Nb_3Sn lines, which indicated a rise in the solubility of tin in the compound Nb_3Sn at this temperature. In alloys Nos. 13, 14, 15, and 16 quenched from 820, 860, and 880° there was a new phase, apparently the intermetallic compound Nb_6Sn_5, in equilibrium with β-Nb_3Sn (alloy No. 13), $NbSn_2$ (alloys Nos. 14, 15, and 16,

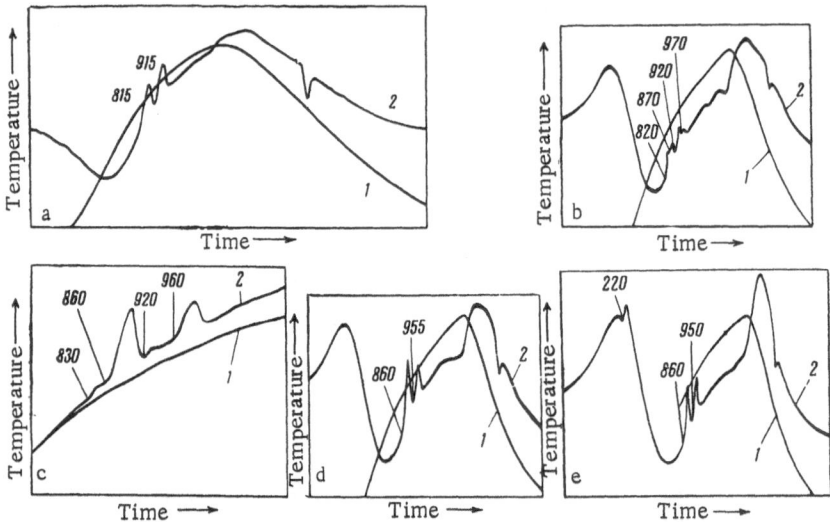

Fig. 2. Thermograms of alloys of the Nb–Sn system. Tin content in at.%: a) 42.9; b) 53.2; c) 60.0; d) 65.0; e) 71.9. 1) Sample temperature; 2) temperature difference between sample and standard.

quenched from 820 to 860°), or with tin (alloy No. 16, quenched from 880°). The compound $NbSn_2$ vanished from all alloys quenched from 880°C or over. In alloys quenched from 930 and 1000°C only two phases appeared: β-Nb_3Sn and tin.

On the basis of these results we concluded that the horizontal at 815°C represented the eutectoid equilibrium $Nb_6Sn_5 \rightleftarrows \beta$-$Nb_3Sn + NbSn_2$, the horizontal at 860° corresponded to the peritectic equilibrium $Nb_6Sn_5 + L \rightleftarrows NbSn_2$, the horizontal at 920° to the peritectic equilibrium β-$Nb_3Sn + L \rightleftarrows Nb_6Sn_5$, and the horizontal at 960° constituted a geometrical representation of the monotectic equilibrium $L_1 \rightleftarrows L_2 + \beta$-$Nb_3Sn$. The existence of a monotectic equilibrium was proposed in [7]; however, the authors of all later papers rejected it on the basis of the low solubility of niobium in molten tin, the experimental shape of the liquidus curve, and so on. Our own data regarding the existence of some nonvariant equilibrium at 960° (characterized by the fact that the phases in equilibrium above and below this temperature were the same) could only be explained on the assumption that this equilibrium was monotectic. In order to reconcile this principle with the results of other authors relating to the low solubility of niobium in molten tin and the shape of the liquidus curve, we had to consider that the peak corresponding to the phase separation of the liquid into $L_1 + L_2$ was very small (about 2% wide and 30° high) and lay almost right up against the tin side of the diagram (see Fig. 1). This latter proposition is supported by the fact that the thermal effect of the monotectic reaction at 960° on the heating thermograms rises with increasing concentration of tin in the alloy. In addition to this, for some alloys (Nos. 12, 17, and 19) thermal analysis was continued to fairly high temperatures (1500 to 1700°C); there was no effects on the thermograms above 960°; the samples showed some signs of melting but were not melted completely.

Summarizing the experimentally-established facts, we came to the conclusion that the Nb–Sn alloys in the equilibrium state contained three intermediate phases, Nb_3Sn, Nb_6Sn_5, and $NbSn_2$, the Nb_3Sn and $NbSn_2$ being stable from their temperatures of formation by a peritectic reaction ($a + L \rightarrow \beta$-Nb_3Sn at 2130 and $Nb_6Sn_5 + L \rightarrow NbSn_2$ at 860°C respectively) to room temperature, while Nb_6Sn_5 only existed within a narrow temperature range; from 920°C, the temperature of formation by the peritectic reaction β-$Nb_3Sn + L \rightarrow Nb_6Sn_5$, to 815°, the temperature of eutectoid decomposition $Nb_6Sn_5 \rightarrow \beta$-$Nb_3Sn + NbSn_2$. At 960°C and about 97% Sn there was apparently a monotectic equilibrium $L_1 \rightleftarrows L_2 + \beta$-$Nb_3Sn$.

For cast Nb–Sn alloys we may apparently construct a metastable equilibrium diagram from which $NbSn_2$ and Nb_6Sn_5 are entirely absent but in which there is a monotectic reaction $L_1 \rightarrow L_2 + Nb_3Sn$ type. Actually the heating thermograms of the cast alloys (heated above 960° and then cooled) only showed the thermal effect of the melting tin at about 230°C and a thermal effect at 850 to 860°C. Since x-ray analysis showed no phases apart from Nb_3Sn and tin in the cast (heated above 960°) alloys, we considered that the effect at 850 to 860° corresponded to the monotectic reaction $Nb_3Sn + L_2' \rightarrow L_1'$. Clearly, in view of the fact that the equilibrium was metastable, the composition of the monotectic alloy and the equilibrium temperature might be different from those of the stable diagram. The diagram of [7] probably constitutes such a metastable diagram, the temperature of the monotectic reaction lying still lower in this case (730°).

The nonmonotonic temperature dependence of the curve representing the solubility of tin in Nb_3Sn is thus due to the fact that at 960° the solubility curve cuts the monotectic horizontal, at 920° the peritectic horizontal, and at 815° the eutectic horizontal. At each of these three points of intersection the solubility curve should experience a break; at one of the breaks the sign of the solubility changes. Apparently the solubility sign change occurs at the intersection with the monotectic horizontal at 960°C for an alloy containing 32% Sn (see Fig. 1).

Conclusions

1. We have plotted the complete phase equilibrium diagram of the Nb–Sn system.

2. We have found that apart from Nb_3Sn the Nb–Sn alloys contain two other intermetallic compounds, Nb_6Sn_5 and $NbSn_2$; these are only formed, however, after prolonged annealing. The compound $NbSn_2$ is formed by a peritectic reaction $Nb_6Sn_5 + L \rightarrow NbSn_2$ at 860°C and is stable down to room temperatures. The compound Nb_6Sn_5 is formed by a peritectic reaction $\beta\text{-}Nb_3Sn + L \rightarrow Nb_6Sn_5$ at 920°C and is stable down to 815°C, where it decomposes in a eutectoid reaction $Nb_6Sn_5 \rightarrow \beta\text{-}Nb_3Sn + NbSn_2$.

3. We consider that at 960°C Nb–Sn alloys experience a monotectic equilibrium $L_1 \rightleftarrows L_2 + \beta\text{-}Nb_3Sn$. The monotectic point lies at about 97% Sn.

4. For cast Nb–Sn alloys in which the compounds Nb_6Sn_5 and $NbSn_2$ are absent but a monotectic reaction $L_1 \rightarrow L_2 + \rightarrow Nb_3Sn$ does take place, we may evidently plot a metastable equilibrium diagram. In view of the fact that the equilibrium is metastable, the temperature of the monotectic equilibrium and the concentration corresponding to the monotectic point may differ from those observed in the stable diagram. A metastable Nb–Sn diagram was probably obtained in [7].

LITERATURE CITED

1. V. N. Svechnikov, V. M. Pan, and Yu. I. Beletskii, Dokl. Akad. Nauk SSSR, 166(6):1328 (1966).
2. J. H. N. Van Vucht, D. J. Van Ooijen, and H. A. C. M. Bruning, Philips Res. Reports, 20:136 (1965).
3. T. R. Anantharaman, Nuclear Science Abstracts, 19(16):30730 (1965).
4. T. H. Ellis and H. A. Wilhelm, J. Less-Comm. Metals, 7:67 (1964).
5. V. N. Svechnikov, Yu. A. Kocherzhinskii, A. K. Shurin, V. M. Pan, A. Ts. Spektor, G. F. Kobzenko, and Yu. A. Boiko, Questions of Metal Physics and Metallography [in Russian], No. 16, Izd. AN Ukr.SSR, Kiev (1962), p. 220.
6. L. J. Vieland, RCA Review, 25:366 (1964).
7. M. I. Agafonov, V. V. Baron, and E. M. Savitskii, Izv. Akad. Nauk SSSR, Otd. Tekh. Nauk, Metallurgiya i Toplivo, No. 5, p. 139 (1959).
8. R. E. Enstrom, G. W. Pearsall, and J. Wulff, J. Metals, 16:97 (1964).

PHASE DIAGRAM OF THE Nb—Ge SYSTEM

V. M. Pan, V. I. Latysheva, and E. A. Shishkin

The phase diagram of the Nb—Ge system is plotted. Three intermetallic compounds are formed in the Nb—Ge system: Nb_3Ge, Nb_5Ge_3, and $NbGe_2$. The compound Nb_3Ge melts with decomposition in a peritectic reaction at 1970°, while Nb_5Ge_3 and $NbGe_2$ melt congruently at 2110 and 1670°C respectively. There are three eutectic equilibria in this system: at 1940, 1590, and 950°C. The ranges of homogeneity of the phases present are determined.

An analysis of the superconducting characteristics of A_3B compounds with the β-W (A15) type of crystal structure formed by niobium with vanadium and elements of the IIIB and IVB subgroups of the Mendeleev periodic system shows that, as a rule, the superconducting characteristics increase as the atomic number and mass number of the B element fall. This relates primarily to the critical temperatures at which these compounds pass into the superconducting state (see Table 1).

The only compound infringing this rule is Nb_3Ge. Up to the present time its value of T_K as determined by various authors [7, 13, 14, 15] has fluctuated between 4.9 and 6.9°K. However, in a recent paper by Matthias et al. [16] it was found that the Nb_3Ge obtained by sharp quenching (cooling rate over 10^6 deg/sec) in the presence of an excess of germanium had a critical temperature of some 17°K. These authors associate this high value of T_K with the fact that the alloy is close to stoichiometric composition (25. at.% Ge). Such results have naturally aroused great interest in the phase diagram of the Nb—Ge system, particularly in the position of the boundaries to the range of homogeneity of the β-Nb_3Ge phase. However, nothing has been published in relation to this phase diagram, although the intermetallic compounds formed in the corresponding alloys have been studied in detail by many authors.

TABLE 1. Critical Temperatures of β-W Type Niobium and Vanadium Compounds

Compound	T_K, °K	Literature cited	Compound	T_K, °K	Literature cited
Nb_3In	9.2	[1]	V_3Sn	3.8	[6]
Nb_3Ga	14.5	[2]	V_3Ge	6.01	[7]
Nb_3Al	17.1—18.0	[2—5]	V_3Si	17.0	[8]
Nb_3Sn	18.1	[9]	V_3In	?	—
Nb_3Ge	6 9—17.0	[13—15.16]	V_3Ga	14.5	[10]
Nb_3Si	—	—	V_3Al	—	—

The compound Nb_3Ge was first studied in [11, 12]. It was found that Nb_3Ge had a crystal structure of the β-W type; the phase based on this compound formed a range of solid solutions extending from $NbGe_{0.22 \pm 0.01}$ to $NbGe_{0.15 \pm 0.01}$ (18 to 14 at.% Ge) at 1600°C, with lattice constants a = 5.167 ± 0.002 and 5.177 ± 0.02 Å, respectively.

Germanides with the formula Me_5Ge_3 were studied in [17]. It was found that the phase of composition $NbGe_{0.54 \pm 0.06}$ (35 at.% Ge) had the same structure as the high-temperature tetragonal form of Nb_5Si_3, which has a structure of the W_5Si_3 (T_1) type. The lattice constants were determined as a = 10.14; c = 5.15 Å, c/a = 0.507. Another compound $NbGe_{0.67 \pm 0.5}$ (40 at.% Ge) was also observed; this had the Mn_5Si_3 [$D8_8$] structure with lattice constants a = 7.71; c = 5.37 Å; c/a = 0.695. This kind of structure is characteristic of silicides of the Me_5Si_3 type containing carbon and oxygen impurities. The niobium used to prepare the alloys contained 0.4% carbon and a little oxygen; the authors accordingly refrained from saying that this phase would remain stable in the absence of carbon and oxygen.

It was shown in [18] that the compound $NbGe_2$ formed in the system under consideration has a hexagonal structure of the $CrSi_2$ type with lattice constants a = 4.967, c = 6.784 Å.

The aim of the present investigation was to plot the phase diagram of the niobium–germanium system.

For this purpose we prepared 26 alloys 10 g each in weight from 99.6% niobium and 99.999% germanium in an arc furnace with a nonconsumable tungsten electrode and a water-cooled copper base in an argon atmosphere. Chemical analysis showed that the losses of germanium in the alloys were negligible, and the composition of the alloys agreed with that of the prepared charge. The compositions of the alloys are given in Table 2.

The cast alloys were studied by microstructural analysis and x-ray diffraction.

Alloys Nos. 1 to 15 were annealed at 1850°C for 5 h and alloys Nos. 1 to 16 and 20 at 1450°C for 100 h, after which they were studied by microstructural analysis and x-ray diffraction. The x-ray pictures were taken with rotating powder samples in a cylindrical camera 57.3 mm in diameter, using unfiltered chromium radiation. Alloys Nos. 5, 13, and 15 were subjected to x-ray analysis after quenching from 1930° (alloy No. 5) and 1950° (alloys Nos. 13 and 15) in castor oil.

All the alloys were studied by differential thermal analysis; for this Nos. 16 to 26 were in the cast state.

TABLE 2. Chemical Compositions of the Alloys Studied

Number of alloy	Germanium content (in the charge)		Weight % of germanium by chemical analysis	Number of alloy	Germanium content (in the charge)		Weight % of germanium by chemical analysis
	at.%	wt.%			at.%	wt.%	
1	5	3.95	3.6	14	38	31.92	—
2	10	7.99	10.7	15	40	34.25	33.4
3	15	12.12	11.9	16	45	39.00	—
4	18	14.64	14.7	17	50	43.86	43.9
5	20	16.34	18.0	18	55	48.85	—
6	22	18.06	18.00	19	60	53.96	52
7	24	19.79		20	67	59.20	—
8	25	20.66	—	21	70	64.58	63.7
9	26	21.54	—	22	75	70.10	—
10	28	23.31	—	23	80	75.76	78
11	30	25.09	22.8	24	85	81.58	—
12	33	28.09	—	25	90	87.55	85.7
13	35	29.61	—	26	95	93.69	—

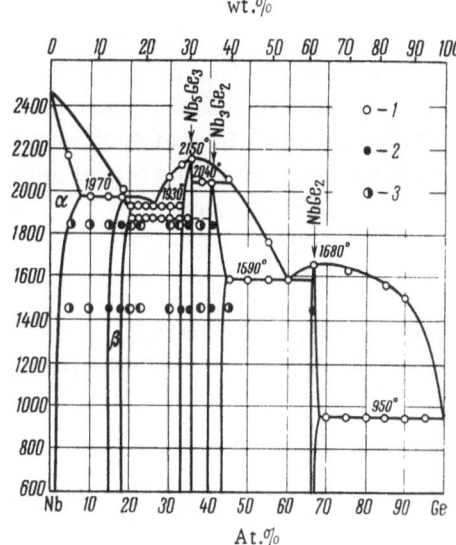

wt.%

Fig. 1. Phase diagram of the Nb–Ge
system. 1) Thermal analysis; 2) one
phase; 3) two phases.

On the basis of these investigations we plotted the phase diagram of the Nb–Ge system; this is shown on Fig. 1. The form and position of the liquidus and solidus curves and also the positions of all the horizontals corresponding to three-phase nonvariant equilibria were determined by differential thermal analysis.

We found that the solid solution of germanium in niobium formed a narrow single-phase region, the limiting solubility of germanium in niobium at the temperature of the peritectic equilibrium (1970° C) being 7 at.%.

The solubility of niobium in germanium was evidently extremely low.

The β phase on the intermetallic compound Nb_3Ge (with the β-W structure) was formed by a peritectic reaction $\alpha + L \rightarrow \beta$-$Nb_3Ge$ at 1970°C. The range of homogeneity of the phase extended in the direction of niobium from the point of stoichiometric composition; it lay between 16 and 21 at.% Ge at 1850 and between 15 and 19 at.% Ge at 1450°C.

X-ray analysis of cast and annealed (at 1850 and 1450°) Nb–Ge alloys revealed three intermediate phases in addition to β-Nb_3Ge. One of these was formed at a concentration of 66.6 at.% Ge and constituted the intermetallic compound $NbGe_2$; the two other intermediate phases were formed at 40 to 43 and 34 to 36 at.% Ge. As yet we have not solved the structure of these compounds and cannot specify what form this may take; we shall provisionally call them Nb_3Ge_2 (40-43 at.% Ge) and Nb_5Ge_3 (34-36 at.% Ge).

The intermetallic compound Nb_5Ge_3 melts congruently at 2150°C. At 1930°C there is a eutectic equilibrium $L \rightleftharpoons \beta$-$Nb_3Ge + Nb_5Ge_3$. The eutectic point lies at 26 to 27 at.% Ge. The thermograms of all the alloys undergoing a eutectic reaction at 1930° showed yet another slight thermal effect on heating, starting at 1870°C. This effect yields no clear peak but forms a bend on the differential curve at 1870°C. This effect appears more sharply in the form of a slight peak on cooling some of the alloys (Nos. 12 and 13). We hypothetically associate the presence of a horizontal on the phase diagram at 1870° with a polymorphic transformation in Nb_5Ge_3. However, no supplementary data supporting this view are as yet available. We were unable to fix the high-temperature form of Nb_5Ge_3 by oil-quenching from 1950°C in an inert medium. We also found no thermal effect at 1870° on the thermogram of alloy No. 14, lying on the other side of the range of homogeneity of Nb_5Ge_3 (in the two-phase $Nb_5Ge_3 + Nb_3Ge_2$ region).

The compound Nb_3Ge_2 has a single-phase region at 40 to 43 at.% Ge and is formed by a peritectic reaction $Nb_5Ge_3 + L \rightarrow Nb_3Ge_2$ at 2040°C.

The intermetallic compound $NbGe_2$ melts congruently at 1680°C. This compound takes part in the following eutectic reactions: with Nb_3Ge_2 at 1590° ($L \rightleftharpoons NbGe_2 + Nb_3Ge_2$) and with germanium at 950° ($L \rightleftharpoons NbGe_2 + Ge$).

Thus our most interesting basic conclusion is that, although the solubility of germanium in β-Nb_3Ge increases with temperature, this rise is very slight (15 to 19 at.% at 1450 and 16 to 21 at.% Ge at 1850°C). The maximum germanium concentration which may be obtained in the β-Nb_3Ge phase is about 21 at.% Ge. Evidently the result obtained by Matthias et al. in [16] in a sharply-quenched alloy is to be explained by the fact that, in this case, the maximum possible germanium concentration in the β phase was secured. We must further suppose that the T_K of the β-Nb$_3$Ge phase depends very sharply on concentration; for a concentration of 18 to 19 at.% Ge it is no greater than 6.9°K, whereas for a concentration of about 21 at.% Ge it rises to almost 17°K (according to Matthias [16]). It is possible that under the conditions of such sharp quenching as that employed by Matthias (cooling rate 10^6 deg/sec) a β phase with a higher germanium concentration than we were able to secure by oil-quenching was achieved.

LITERATURE CITED

1.   M. D. Banus, T. B. Reed, and H. C. Gatos, J. Phys. Chem. Solids, 23:971 (1962).
2.   E. A. Wood, V. B. Compton, B. T. Matthias, and E. Corenzwit, Acta Cryst., 11:604 (1958).
3.   E. Corenzwit, J. Phys. Chem. Solids, 9:93 (1959).
4.   K. Raetz and E. Z. Saur, Physik, 169:315 (1962).
5.   P. S. Schwartz, Phys. Rev. Lett., 9:448 (1962).
6.   G. D. Cody, J. J. Hanak, G. T. McConville, and F. D. Rosi, Phys. Rev. Lett., 6:275 (1961).
7.   G. F. Hardy and J. K. Hulm, Phsy. Rev., 93:1004 (1954).
8.   N. E. Alekseevskii, E. M. Savitskii, V. V. Baron, and Yu. V. Efimov, Dokl. Akad. Nauk SSSR, 145:82 (1962).
9.   B. T. Matthias, Phys. Rev., 97:74 (1955).
10.   J. N. Wernick, F. J. Morin, S. L. Hsu, D. Dorsi, J. P. Maita, and J. E. Kunzler, Phys. Rev. Lettl, 5:149 (1960).
11.   J. H. Carpenter and A. W. Searcy, J. Am. Chem. Soc., 78:2079 (1956).
12.   J. H. Carpenter, J. Phys. Chem., 67:2141 (1963).
13.   T. B. Reed, H. C. Gatos, W. J. La Fleur, and J. T. Roddy, Metallurgy of Advanced Electronic Materials (ed. G. E. Brock), Vol. 19 (1963), p. 71.
14.   B. T. Matthias, T. H. Geballe, and V. B. Compton, Rev. Mod. Phys., 35:1 (1963).
15.   C. T. Thompson and J. K. Gerber, Solid State Electronics, 2:259 (1961).
16.   B. T. Matthias, T. H. Geballe, R. H. Willens, E. Corenzwit, and G. W. Hull, Phys. Rev., 139:A1501 (1965).
17.   H. Nowotny, A. W. Searcy, and J. E. Orr, J. Phys. Chem., 60:677 (1956).
18.   J. H. Carpenter and A. W. Searcy, J. Phys. Chem., 67:2144 (1963).

PHASE DIAGRAM OF THE Nb—Nb$_5$Si$_3$ SYSTEM
AND CRYSTAL STRUCTURE OF THE COMPOUND Nb$_3$Si

V. M. Pan, V. V. Pet'kov, and O. G. Kulik

The phase diagram of the Nb–Nb$_5$Si$_3$ system is presented. In alloys of this system there is a compound Nb$_3$Si formed by a peritectic reaction at 1880°C; it is stable down to 1800°. Prolonged annealing at lower temperatures leads to the decomposition of this compound. The critical structure of Nb$_3$Si is determined (Ti$_3$P type, space group P4$_2$/n-C$_{4r}^4$, tetragonal cell, a = 10.230 Å, c = 5.189 Å, c/a = 0.507).

The most "high-temperature" and high-field superconductors known at the present time are compounds of the A$_3$B type with the β-tungsten structure. The highest critical superconducting temperature is that of Nb$_3$Sn: 18.05°K [1]. Intensive searches for materials with higher values of T$_K$ have not so far been successful.

Recently certain research workers [2, 3] suggested that if there were such a compound as Nb$_3$Si with the β-tungsten structure then this would have a critical temperature much higher than 18°K. This has aroused great interest in the Nb–Si system and the corresponding phase diagram.

The phase diagram of the Nb–Si system was first plotted by Knapton [4] and then by Kieffer et al. [5] and also G. V. Samsonov et al. [6]. The authors of [4, 6] observed three intermediate phases in the alloys of this system: NbSi$_2$, Nb$_5$Si$_3$ and Nb$_4$Si. The last of these was not found in [5]. S. I. Alyamovskii et al. [7] also only found two intermediate phases, NbSi$_2$ and Nb$_5$Si$_3$ at temperatures up to 1600°C. P. M. Arzhanyi et al. [8] found a compound Nb$_4$Si in diffusion layers on silicizing niobium between 1100 and 1300°C. Goldschmidt [9] suggested that Nb$_4$Si was a high-temperature phase.

Recently reports in which Nb$_3$Si replaces Nb$_4$Si have appeared. These include papers by Galasso and Pyle [2] and Rossteutscher and Schubert [10]. However, the authors of [2] obtained this compound by annealing at 1100°C and the authors of [10] only by annealing at higher temperatures (1800°C).

Just as contradictory are the results of different authors regarding the melting point of the niobium-richest compound of the Nb–Si system, and in particular its crystal structure. According to Knapton [4], Nb$_4$Si is formed by a peritectic reaction at about 1950°C and is isomorphic with Ta$_4$Si and Zr$_4$Si, while according to [6] it melts congruently at about 2580°C and has a hexagonal cell of the ε-Fe$_3$N, UCl$_3$ or BCl$_3$ type with lattice constants a = 3.59 and c = 4.46 Å. The authors of [2]* consider that the crystal lattice of Nb$_3$Si (and not Nb$_4$Si) belongs to the cubic system, and the arrangement of the atoms is the same as in the ordered Cu$_3$Au structure; the lattice constant is a = 4.211 Å. Finally, in [10] the structure of Nb$_3$Si is described as te-

---

*It was found in [2] that Nb$_3$Si passed into the superconducting state at 1.5°K.

tragonal, isomorphic with Ta_3Si and belonging to the Ti_3P type; the lattice constants of Nb_3Si are a = 10.23, c = 5.19 Å, c/a = 0.507.

We started the present investigation in order to refine the phase diagram of the high-niobium part of the Nb—Si system (particularly the range of concentrations close to 4Nb1Si and 3Nb1Si) and also in order to determine the crystal structure of the compound Nb_4Si (Nb_3Si).

The original materials for preparing the alloys were niobium in the form of short rods (99.6%) and silicon of the single-crystal type known as KSITD-11. The alloys were prepared in an arc furnace with a tungsten electrode and a water-cooled copper base in an argon atmosphere. No chemical analysis of the alloys was made. The composition was checked by weighing the solidified material after melting. The weight of the solid bar differed very little from the sum of the weights of the constituents. It was therefore considered that the composition of the alloy corresponded to that specified in the charge. Table 1 shows the composition of the alloys studied.

The alloys were studied by differential thermal, x-ray-structural, and microstructural analyses. Thermal analysis was carried out in hafnium dioxide crucibles (thermocouple W—W 20%Re) in an atmosphere of purified argon. The x-ray diffraction pictures were taken from rotating powder samples in cylindrical cameras 57.3 and 114.6 mm in diameter in unfiltered chromium radiation. The alloys were annealed in a TVV-4 furnace in an argon atmosphere.

In the cast state the alloys were studied by microstructural analysis and x-ray diffraction. The microstructure of the cast samples containing 20 to 25 at.% Si showed three phases. The x-ray pictures of the cast samples contained a number of lines not belonging to the α phase (Nb-base solid solution) and Nb_5Si_3. Differential thermal analysis showed that the majority of the alloys started melting at 1800°C.

After annealing at 1830° for 40 h, the alloy containing 25% Si showed only the lines of the new phase; the lines of the α solution and Nb_5Si_3 had vanished. The x-ray pictures of alloys with less than 25% Si also showed the lines of the α solution, while those of alloys containing over 25% Si gave the lines of Nb_5Si_3. This provisionally suggested that the new phase was Nb_3Si.

In view of the fact that it was extremely difficult to prepare microsections of alloys annealed at 1830° and containing a large amount of the new phase, no microstructural analysis of these samples was carried out. After annealing at 1400° for 78 h the x-ray pictures of all the alloys showed only the lines of the α solution and Nb_5Si_3. On varying the composition, only the intensity ratio of the lines of these two phases varied.

Then we made a differential thermal analysis of the alloys annealed at 1400°C. From the results of the analysis we plotted the melting diagram of Fig. 1. We found that between Nb and Nb_5Si_3 two nonvariant equilibria involving the liquid phase occurred in the Nb—Si system: a eutectic L $\rightleftarrows$ α + Nb_5Si_3 at 1880 + 10°C and a peritectic Nb_5Si_3 + L $\rightleftarrows$ Nb_3Si at 1920 + 10°C. The eutectic point lay at 18 at.% Si. Our melting diagram agreed closely with Knapton's results, [4].

TABLE 1. Composition of the Alloys Studied

Number of alloy	Silicon content		Number of alloy	Silicon content	
	at.%	wt.%		at.%	wt.%
1	5	1.57	5	25	9.16
2	10	3.25	6	30	11.47
3	15	5.07	7	35	13.99
4	20	7.03	8	37.5	15.36

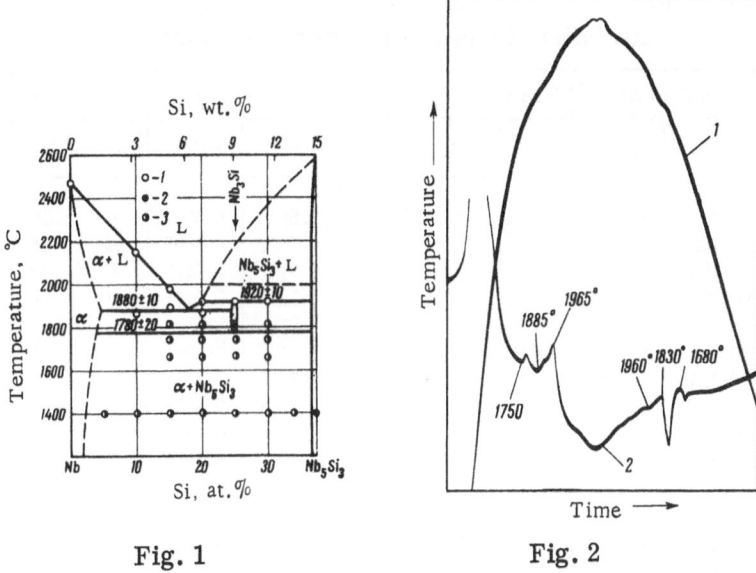

Fig. 1   Fig. 2

Fig. 1. Phase diagram of the Nb–Nb$_5$Si$_3$ system. 1) Thermal analysis; 2) one phase; 3) two phases.

Fig. 2. Thermogram of Nb–15%Si. 1) Sample temperature; 2) difference between the temperatures of the sample and the standard.

Figure 2 shows, for example, the thermogram of the alloy containing 15 at.% Si. Here the onset of melting (eutectic reaction) occurs at 1885° and the end of melting at 1965°C; the onset of crystallization occurs at 1960° and the crystallization of the eutectic at 1830°C. The effects at 1750° (heating) and 1680° (cooling) are associated with the polymorphism of the HfO$_2$ crucible material.

No thermal effect corresponding to the formation of Nb$_3$Si from the two-phase mixture α + Nb$_5$Si$_3$ on heating appears on the thermograms. This is understandable. Only prolonged annealing at 1830°C leads to the formation of this phase. Such a slow process cannot take place on heating the sample in a thermal-analysis apparatus at a rate of 50 to 100 deg/min or over. On cooling the alloy containing 25 at.% Si the differential curve shows a weak thermal effect at about 1600°C, which may be associated with the decomposition of Nb$_3$Si formed by peritectic crystallization. It should be noted that peritectic crystallization is largely suppressed on rapid cooling, and the thermograms representing the cooling of the alloys with 25 and 30 at.% Si show an effect due to the crystallization of a nonequilibrium eutectic. It is clear that the peritectic reaction may be almost entirely suppressed in the case of a fairly high cooling rate, and then the transformation taking place in the Nb–Nb$_5$Si$_3$ alloys may only be described by a metastable-equilibrium diagram with a eutectic horizontal. It was probably a diagram of this kind which Kieffer et al. plotted in [5].

The thermograms recorded during the cooling of alloys with 25% Si or over show a sharp thermal effect at about 1500°C. This effect may possibly be explained by a polymorphic transformation in the compound Nb$_5$Si$_3$, as described in [4-6]. According to the published data, this transformation occurs at 1900 to 2000°C. However, our own heating diagrams fail to reveal the corresponding effect at these or any other temperatures, while the effect on the cooling thermograms may be attributed to a transformation of material severely supercooled from the equilibrium temperature. For this reason we show the 2000° Nb$_5$Si$_3$ polymorphism by a broken line in Fig. 1.

TABLE 2. Powder X-Ray Diffraction Record of Nb_3Si* (Chromium Radiation, Camera Diameter 114.6 mm)

hkl	Value of $\sin^2 \vartheta$		Line intensity†		hkl	Value of $\sin^2 \vartheta$		Line intensity †	
	expt.	calc.	obs.	calc.		expt.	calc.	obs.	calc.
201	0.0984	0.0988	W	23.8	622	0.6972	0.6967	W	16.4
221	0.1425	0.1489	W	13.9	641	0.7022	0.7007	W	16.5
301	0.1605	0.1617	W	8.47	721	0.7101	0.7132	W	10.3
311	0.1726	0.1742	M	28.0	730	0.7290	0.7271	W	4.0
002	0.1932	0.1952	W	9.5	433⎫	0.7539	0.7525	W	4.6
102	0.2058	0.2077	W	5.6	503⎭				
321	0.2137	0.2117	VS	100.0	513	0.7655	0.7642	W	5.5
112	0.2192	0.2203	S	120.0	731	0.7769	0.7759	W	5.6
330	0.2251	0.2257	S	81.5	004	0.7799	0.7797	M	35.2
202	0.2447	0.2453	M	28.4	104	0.7939	0.7922	M	14.0
401⎫	0.2490	0.2493	M	48.6	523⎫	0.8004	0.8017	W	3.4
420⎭		0.2506			800⎭				
411	0.2601	0.2619	VS	112.0	702	0.8084	0.8015	W	1.9
331	0.2730	0.2745	VW	0.2	651	0.8185	0.8085	VW	3.1
222	0.2937	0.2955	M	62.1	552⎫	0.8214	0.8136	VS	83.0
421	0.2996	0.2995	M	27.4	712⎭		0.8211		
302	0.3078	0.3080	W	18.4	204	0.8325	0.8298	W	1.1
510	0.3257	0.3259	M	34.2	642	0.8471	0.8462	M	40.0
511	0.3733	0.3748	W	6.2	820⎫	0.8540	0.8516	W	1.5
402	0.3960	0.3958	VW	2.9	722⎭		0.8587		
621	0.5500	0.5503	W	1.0	533	0.8651	0.8643	W	1.7
313	0.5633	0.5644	W	4.6	224	0.8830	0.8799	W	2.8
442	0.5968	0.5964	M	18.8	304	0.8926	0.8924	W	2.8
323	0.6028	0.6020	M	9.0	314⎫	0.9027	0.9049	M	30.4
631	0.6139	0.6130	W	7.4	613⎭		0.9019		
532	0.6198	0.6215	M	0.1	732	0.9222	0.9213	W	12.2
710⎫	0.6274	0.6260	M	35.3	750	0.9283	0.9267	M	21.4
550⎭					623	0.9398	0.9395	VW	0.8
403	0.6426	0.6396	W	12.7	543⎫	0.9514	0.9520	W	22.8
602	0.6442	0.6465	M	32.5	661⎭		0.9504		
413	0.6524	0.6522	S	28.0	652	0.9596	0.9589	VW	1.6
640		0.6519			831	0.9642	0.9629	S	74.2
711⎫	0.6755	0.6756	W	6.4	751	0.9771	0.9754	S	61.2
551⎭					414	0.9927	0.9926	S	78.5
423	0.6911	0.6898	W	3.3					

* Only the α line of Nb_3Si given.

† W = weak, VW = very weak, S = strong, VS = very strong, M = medium.

In order to determine the temperature of the eutectoid equilibrium $Nb_3Si \rightleftarrows \alpha + Nb_5Si_3$ we carried out a series of annealings at 1600 (50 h), 1680 (96 h), and 1750° (50 h). We found that at all these temperatures alloys with 15 to 30 at.% Si retained the phase composition $\alpha + Nb_5Si_3$. The compound Nb_3Si was only formed on annealing at 1820 to 1830°C (Schubert [10] claims it to be formed at 1800° also). We concluded that the temperature of the eutectoid equilibrium $Nb_3Si \rightleftarrows \alpha + Nb_5Si_3$ was about 1780°C.

We also determined the crystal structure of Nb_3Si. Powder samples of the alloy with 25 at.% Si were annealed at 1830°C for 40 h and exposed in an RKU-114M camera in unfiltered chromium radiation for 15 h. The results are shown in Table 2.

We found that all the lines on this x-ray picture could be indexed on the assumption of a tetragonal lattice with parameters a = 10.230, c = 5.189 Å, c/a = 0.507. The squares of the sines of the experimental and calculated reflection angles agreed closely. Attempts to index the lines of the Nb_3Si x-ray diffraction photograph on the basis of the results given in [2, 4, 6] yield no satisfactory conclusion.

We calculated the line intensities on the assumption that Nb_3Si had a structure of the Ti_3P type and that the niobium and silicon atoms occupied positions with the same coordinates as tantalum and silicon in Ta_3Si [10]. The observed intensities agreed quite well with the calculated values (Table 2) for the following positions of the atoms:

$$3 \times 8Nb \text{ in } (g): \begin{array}{ccc} 0.037 & 0.560 & 0.237 \\ 0.148 & 0.665 & 0.719 \\ 0.105 & 0.231 & 0.529 \end{array}$$
$$8Si \text{ in } (g) \quad : 0.041 \quad 0.231 \quad 0.013$$

The unit cell of Nb_3Si contains 32 atoms (24 of niobium and 8 of silicon). In calculating the intensities the following factors were taken into account: structure, angles, and multiplicity.

Analysis of the results leads to the conclusion that Nb_3Si belongs to the Ti_3P type of structure (space group $P4_2/n - C_{4h}^4$) and has a tetragonal cell with lattice constants a = 10.230, c = 5.189 Å, c/a = 0.507.

We also indexed the x-ray diffraction picture of the low-temperature form of the compound Nb_5Si_3. We confirmed that this form of Nb_5Si_3 belonged to the Cr_5B_3 type of structure (space group $14/mcm - D_{4h}^{18}$) and had a tetragonal cell with lattice constants a = 6.57, c = 11.884 Å, c/a = 1.808.

Conclusions

1. We have plotted the phase diagram of the $Nb-Nb_5Si_3$ system.

2. We have found that this system contains an intermetallic compound Nb_3Si, which under equilibrium conditions is formed by a peritectic reaction $Nb_5Si_3 + L \rightleftarrows Nb_3Si$ at 1920°C and decomposes by a way of a eutectic reaction $Nb_3Si \rightleftarrows \alpha + Nb_5Si_3$ at about 1780°C.

3. We have shown that the compound Nb_3Si belongs to the Ti_3P type of structure (space group $P4_2/n - C_{4h}^4$) and has a tetragonal unit cell with lattice constants a = 10.230, c = 5.189 Å, c/a = 0.507, in agreement with the results of [10].

LITERATURE CITED

1. B. T. Matthias, T. H. Geballe, S. Geller, and E. Corenzwit, Phys. Rev., 95:1435 (1954).
2. F. Galasso and J. Pyle, Acta Cryst., 16:228 (1964).
3. L. Gold, Phys. Stat. Solidi, 4:261 (1964).
4. A. Knapton, Nature, 175:730 (1955).
5. E. Kieffer, F. Benesowsky, and H. Schmidt, Z. Metallkunde, 47:247 (1956).
6. G. V. Samsonov, V. S. Neshpor, and V. A. Ermakova, Zh. Neorg. Khim., 3:868 (1958).
7. S. I. Alyamovskii, P. V. Gel'd, and I. I. Matveenko, Zh. Neorg. Khim., 7:836 (1962).
8. P. M. Arzhanyi, L. M. Volkova, and D. A. Prokoshkin, Izv. Akad. Nauk SSSR, Otd. Tekh. Nauk, Metallurgiya i Toplivo, 6:127 (1959).
9. H. J. Goldschmidt, J. Iron Steel Inst., 194:169 (1960).
10. W. R. Rossteutscher and K. Schubert, Z. Metallkunde, 11:813 (1965).

VI. METHODS OF STUDYING
THE PROPERTIES OF SUPERCONDUCTING ALLOYS

APPARATUS FOR MEASURING THE TEMPERATURE CORRESPONDING TO THE TRANSFORMATION OF METALS AND ALLOYS INTO THE SUPERCONDUCTING STATE

N. D. Kozlova, Yu. V. Efimov, V. V. Baron, and E. M. Savitskii

A method of measuring critical temperatures from 4.2 to 20°K is described; it is based on a special new apparatus for measuring the magnetic permeability of samples and is accurate to 0.2°K. The measuring method is simple and gives reliable and reproducible results even if the amount of superconducting phase in the samples is under $0.5 \cdot 10^{-4}$ cm^3. In measuring alloys containing two or several superconducting phases, the transitional curves give characteristic steps; the temperatures of these steps correspond to the T_K of the phases present. The method may also be used for the phase analysis of superconducting alloys.

It is well known that superconductors are distinguished from other metals by two properties: the absence of electrical resistance, and the expulsion of a magnetic field from the superconducting volume (Meissner effect). Hence the principal methods of determining the transformation temperature are electrical and magnetic [1].

The electrical method lies in measuring the electrical resistance of the sample as a current passes through it; the vanishing of the resistance at some particular temperature indicates transformation into the superconducting state. The magnetic method is based on the Meissner effect; as the superconductor passes into the superconducting state the magnetic field leaves it and its magnetic characteristics (susceptibility, magnetic moment, etc.) alter.

Magnetic methods have a number of advantages over electrical. Firstly, in the electrical method the width of the superconducting transformation depends substantially on the value of the measuring current; it increases with rising current, and this reduces the accuracy of determining the critical temperature. Secondly, magnetic methods enable us to judge the volume properties of the sample, whereas with electrical methods we can only measure the electrical resistance of the sample in the region at which it falls to a minimum. This is particularly important for superconducting alloys; the electrical resistance of the sample may be still zero when magnetic measurements show that only a small proportion of the volume is superconducting. Thirdly, magnetic measurements enable us to determine the cri-

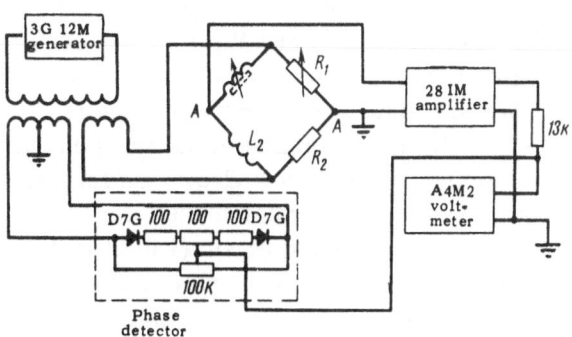

Fig. 1. Electrical circuit of an apparatus for measuring the critical superconducting temperature.

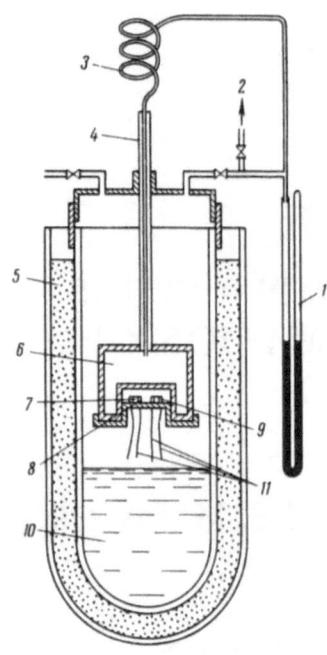

Fig. 2. Schematic representation of the gas thermometer. 1) Manometer; 2) to backing pump; 3) capillary; 4) movable rod; 5) liquid nitrogen 6) gas thermometer; 7) sample; 8) measuring coil; 9) compensating coil; 10) liquid helium; 11) leads.

tical temperature in samples of any arbitrary shape, whereas electrical measurements demand wires.

We ourselves determined the critical temperature by reference to changes in the magnetic permeability of the sample. The transformation was established by means of an inductance bridge (Fig. 1). An ac signal (frequency 1 kc/sec) is applied to a bridge comprising inductances L_1 and L_2 and resistances R_1 and R_2. From the points AA the voltage passes through an amplifier to a dc voltmeter, to which a signal is also fed from a phase detector. A dc voltage is thus received by the voltmeter, and the value of this is proportional to the signal taken from the bridge.

The sample is placed in one of the induction coils (L_1). When the sample passes into the normal state the magnetic permeability changes from $\mu = 0$ to 1 over a certain temperature range. The induction $L_1 \sim \mu$, so that the value of the signal taken from the bridge will be proportional to μ. When the sample is in the superconducting state, the bridge is compensated by means of the resistance R_1. When the sample passes into the normal state, a mismatch signal appears. In our case, for a peak generator ac voltage of 20 V, the mismatch signal rises from 0 to roughly 2 V, which is quite easily observed.

The apparatus is based on a gas thermometer (Fig. 2). The samples are placed within a copper bomb in a copper holder. The temperature is varied by lowering the level of the helium by its evaporation. The rate of temperature change is about 0.15 deg/min. It is easy to calculate that the relative error in determining the temperature $\Delta T/T_x$ equals

$$\frac{\Delta T}{T_x} = \frac{V_b}{V_b + \frac{4.2}{300} V_{ds}\left(1 - \frac{P_x}{P_{4.2}}\right)} - 1,$$

where T_x is the temperature being determined, V_σ is the volume of the bomb, V_{ds} is the dead space (volume of gas in the capillary and manometer, which have $T > T_x$); P_x is the pressure at the temperature T_x; $P_{4.2}$ is the pressure at temperature 4.2°K.

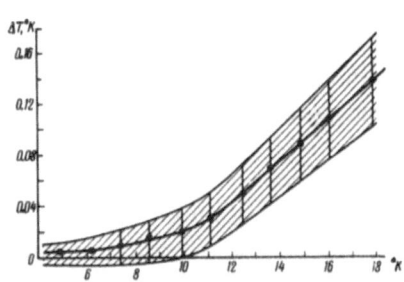

Fig. 3. Systematic error in measuring the temperature in the gas thermometer as a function of temperature.

Figure 3 shows the relation $\Delta T (T_x)$ for $P_{4.2}$ = 170 mm Hg, $V_\sigma = 47$ (± 8) cm^3, and $V_{ds} = 6$ (± 2) cm^3. The shaded region corresponds to the calculated spread (scatter) ΔT. We see that the greatest deviation is about 0.2°K at T = 17.8°K. This kind of accuracy is quite sufficient for studying the T_K of new alloys and investigating macroscopic laws. The experimental error lies within the limits of the calculated error (1 to 2%); this may be seen from the transformation curves of Pb and V–Ga alloys obtained at various times (Fig. 4).

In its first form the apparatus was designed for measuring six samples with one filling of helium. After adoption and development, the number of measured sam-

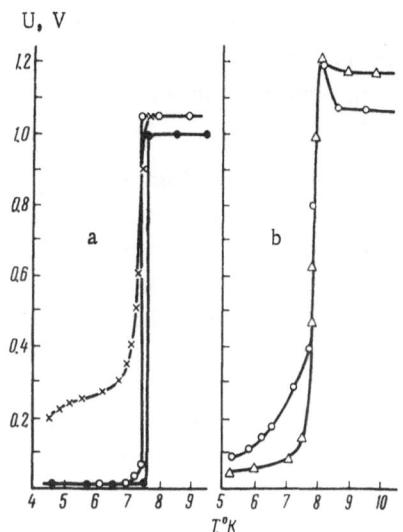

Fig. 4. Critical curves of lead (a) and a V–23%Ga alloy (b) taken at various times.

ples was raised to ten. The induction coils are wound on paper frames (internal diameter 4 mm, height 5 to 6 mm) from PÉL-0.15 conduit. The number of turns on each coil is about 500. The apparatus and the method chosen for measuring the critical temperature are quite simple and give excellently reproducible results.

As indicated in the foregoing, this method of measuring the critical temperature is sensitive to the amount of superconducting phase in the whole volume of the test sample. When determining the temperature of the transformation from the superconducting to the normal state for diffusion V–Ga and V–Si coatings of various thicknesses (1 to 100 μ), critical curves with different values of signal (depending on the amount of superconducting phase of the same composition present) were obtained.

The samples for measuring the T_K of coatings constitute bundles of 10 to 20 wires 0.2 mm in diameter and 10 to 12 mm long. The volume of the superconducting V–Ga or V–Si phases in this kind of bundle, calculated from the formula $V = n \cdot h \ 2\pi r \cdot l$ (n is the number of wires in the sample, h the thickness of the coating, r the radius and l the length of the wires), is $(0.63 \text{ to } 151) \cdot 10^{-4}$ cm^3 for a total sample volume of $(31 \text{ to } 300) \cdot 10^{-4}$ cm^3, i.e., the proportion of superconducting phases in the samples is between 2 and 50% of the total volume. The vanadium core becomes a nonsuperconducting phase at temperatures above 5.3°K. Even with a minimum thickness of the coatings (1 μ) on the samples, for which the volume of the superconducting phase is only 2% of the total volume of the samples, the T_K of the coatings could be measured quite sharply and (within the limits of experimental error) reproducible results could be achieved. With increasing thickness of the coating the value of the signal corresponding to samples of the same phase composition increased.

The sensitivity of the apparatus to the volume of the superconducting phase was verified on specially-prepared samples of identical shape, containing various volume proportions of the same superconducting phase. The standard was a sample of electron-beam-melted niobium 2.5 mm in diameter and 6 mm long $(V = 270 \cdot 10^{-4}$ cm$^3)$. For the measurements we took samples of the same niobium constituting 2.10 and 50% of the volume of the original sample. These samples were placed in copper blocks in the core of the measuring coils.

The value of the signal taken from the bridge for a sample $5.4 \cdot 10^{-4}$ cm^3 in volume (2% of the original) was 0.07 V. The minimum signal measured on the apparatus was 0.01 V. Hence it was possible to measure the T_K of samples containing a still smaller proportion of superconducting phase. In addition to this, by raising the input signal to 20 V the sensitivity of the apparatus could be greatly increased.

With increasing amount of superconducting phase the value of the signal taken from the bridge increases (Fig. 5). For a proportion of superconducting phase exceeding 5% of the total volume of the sample, there is a rectilinear relationship. This relationship is infringed for small quantities of the superconducting phase. This is evidently associated with the experimental error in determining the critical temperature and calculating the volume of the niobium.

The experiment under consideration was carried out with samples of compact niobium of the volume indicated. In the alloys the superconducting (or any other) phase was usually distributed throughout the whole volume of the samples. In order to simulate the alloys, samples $270 \cdot 10^{-4}$ cm^3 in volume were made from fine copper and niobium powder. The amount of nio-

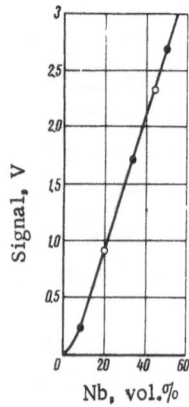

Fig. 5. Value of the signal as a function of the volume proportion of niobium in the samples.

bium in the volume corresponded to 20 and 40%. The values of the signals corresponding to these samples fell almost exactly on the graph representing the relationship between the signals and the volume proportion of the superconducting phase (see Fig. 5).

Thus the foregoing method of measuring the critical temperature may be used for quantitatively determining the volume proportion of the superconducting phases in the alloys. Since the value of the output signal depends not only on the proportion of the phase in the superconducting state but also on the residual resistance of this phase in the normal state, the volume dependence of the signal will not be quite so simple if there are several phases in the sample. Additional calibration graphs will be required in order to determine the volume proportion of the phase. This must be remembered when passing from one system of superconducting alloys to another.

We note that the accuracy of determining the volume proportion of of the superconducting phase in the alloy can hardly be greater than 5 to 10%.

A measurement of the critical temperature may thus serve as a method of physicochemical analysis for studying superconducting alloys [3]. In systems of two superconducting metals or compounds a smooth change in critical temperature occurs when a continuous series of solid solutions is formed [4-6]. At the boundary of each phase transformation there is a sharp change in the form of the curve relating the critical temperature to the composition of the alloys. Examples include the Nb-Ti, V-Ti, and W-Os systems [4, 7]. There is an indication that the changes in the superconducting characteristics relative to the composition of the alloys are mainly associated with changes in solubility; they remain more constant in retions of eutectic alloys [7]. In alloys having small quantities of other superconducting phases with different critical temperatures in addition to the main superconducting phase, the curves are diffuse. This kind of effect also occurs in completely equilibrium alloys of the V-Si and V-Ga systems. If the alloys contain several phases in comparable quantities, the critical curves show bends or in some cases characteristic steps. The temperature of these steps corresponds to the critical temperatures of the phase established in single-phase V-Ga alloys.

In order to verify this situation we pressed samples from uniform mixtures of lead, niobium, and vanadium powders, each metal occupying a specified volume. The critical curves of these samples clearly showed steps corresponding to the critical temperatures of the pure metals (Fig. 6). The critical curves of the mixed samples in no fundamental way differed from those of multiphased V-Ga alloys. The latter simply had rather less sharply-expressed steps, presumably as a result of a certain microinhomogeneity in the alloys.

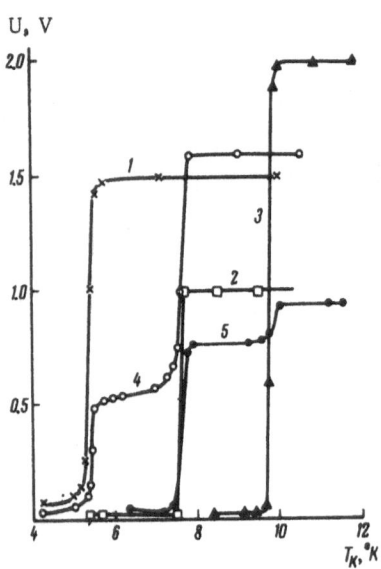

Fig. 6. Transformation curves of vanadium, lead, niobium, and pressed mixtures of the powders of these metals. 1) Electrolytic vanadium; 2) lead; 3) electron-beam-melted niobium; 4) 40% V and 60% Pb (mixture); 5) 20% Nb and 80% Pb (mixture).

Conclusions

We have developed a method of measuring the critical temperature between 4.2 and 20°K and have constructed a special apparatus for this purpose. The accuracy of measuring the superconducting transformation temperature is about 0.2°K. The method is simple and gives a high reproducibility of the results. In measuring the critical temperature reliable results may be obtained even if the samples contain under $0.5 \cdot 10^{-4}$ cm^3 of the superconducting phase. In measuring the critical temperature of alloys in which two or more phases are superconducting, the critical curves show characteristic steps, the temperatures of which correspond to the critical temperatures of the phases present. As a result of this the method in question may also be used for the phase analysis of superconducting alloys.

LITERATURE CITED

1. E. A. Lyton, Superconductivity, Wiley, New York (1962).
2. G. K. White, Experimental Techniques in Low-Temperature Physics, Oxford University Press (1959).
3. E. M. Savitskii and V. V. Baron, Izv. Akad. Nauk SSSR, Metallurgiya i Gornoe Delo, No. 5, 3 (1963).
4. E. M. Savitskii, V. V. Baron, Yu. V. Efimov, V. R. Karasik, T. V. Vylegzhanina, and E. I. Gladyshevskii, Zh. Neorg. Khim., 9(8):2045 (1964).
5. E. M. Savitskii, V. V. Baron, Yu. V. Efimov, and E. I. Gladyshevskii, Izv. Akad. Nauk SSSR, Neorg. Mat., 1(2):208 (1965).
6. J. K. Hulm and R. D. Blaugher, Phys. Rev., 123(5):1569 (1961).
7. M. Tannenbaum and W. V. Wright, J. Metals, 14(5):367 (1962).

USE OF PHYSICAL METHODS FOR REVEALING INHOMOGENEITIES IN THE COMPOSITION AND STRUCTURE OF Zr—Nb ALLOYS

Yu. F. Bychkov, M. T. Zuev, N. A. Sokolov, I. A. Baranov, and R. S. Shmulevich

The nonuniformity in the distribution of Zr and Nb at various points of an Nb—Zr ingot and forged alloy bars is determined to an accuracy of 0.35% by an x-ray method. A camera attachment is devised for continuously monitoring the homogeneity of the structure of a fine wire without damaging the latter. The methods of measuring the thermo-emf and electrical resistance at liquid-nitrogen temperature are extremely suitable for revealing inhomogeneities of the structure in wires subjected to heat treatment. These are very sensitive and well-adapted to the continuous monitoring of long pieces of wire. The results of wire-homogeneity measurements carried out in this way are compared with the degradation of the critical current in solenoids and with critical-current measurements made along the whole length of the wire.

In order to ensure the efficient use of superconductors the technology of their preparation and treatment should guarantee the production of material with specific and stable superconducting properties. The serious degradation in current often observed on passing from short samples to solenoids is undoubtedly associated not only with the specific characteristics of the solenoid tests but with inhomogeneities in the characteristics along the length of the wire. The fact that the critical current density may be inconstant along the length of a superconducting wire was mentioned in [1]. The authors observed a considerable fall in critical current along a heat-treated Zr—50%Nb wire. Investigations showed that the characteristics of the superconducting wires were considerably affected by such metallurgical factors as the composition and structure of the superconductor, the presence of internal stresses, and so on. Hence in order to ensure that a wire should have specific and stable superconducting parameters it is essential that it should be reasonably uniform in structure and composition along its whole length. The existence of inhomogeneous parts arising from the poor quality of the alloy or as a result of heat treatment of the wire and variations in the superconducting characteristics of these may seriously reduce the maximum magnetic field attainable in a solenoid made of such wire. Hence the question of estimating the quality of bars and wires and the technology of preparing and heat-treating these in order to obtain a homogeneous material with specified superconducting characteristics, constant along its whole length, are extremely important.

In this paper we consider certain methods of revealing inhomogeneities in the composition and structure of Zr—Nb bars and wires by physical means.

Checking the Homogeneity of Zr—Nb Bars

In the production of superconducting wire the most important problem is that of obtaining bars of uniform composition.

The homogeneity of Zr–Nb bars prepared by arc or electron-beam melting may arise from the fact that the original metals have considerably different melting points (1860 and 2450°C), different specific gravities (6.5 and 8.6 g/cm³) and different vapor tensions. Inhomogeneity in the composition of a Zr–Nb wire prepared from an inhomogeneous bar leads not only to a fall in the critical current but also to a change in the electrical magnetic field; for example, a change in the niobium content from 75 to 85% in some part of a wire not subjected to heat treatment reduces the critical field at 1°K from 105 to 75 kOe [2]. On subsequent (even ideal) heat treatment of such a wire, the inhomogeneity of composition may lead to structural inhomogeneity and to a quantitative or even qualitative difference in phase composition in individual parts of the wire, since the stability of the original β solid solution varies with varying niobium content.

The main methods of checking the homogeneity of Zr–Nb bars are chemical analysis and measurements of hardness. However, the hardness, as a nonlinear function of concentration, is not sufficiently sensitive a property, while the use of chemical analyses is usually limited and the time taken to carry them out often interferes with the whole flow of the process.

We have verified the possibility of determining the composition of Zr–Nb bars and ingots very accurately by an x-ray method based on the lattice constant of the solid solution. The possibility of using the x-ray method is based on the fact that in the cast state zirconium alloys containing over 15% Nb* have the structure of a β solid solution in which the bcc lattice constant varies sharply and linearly with Nb content (Fig. 1).

X-ray diffraction pictures were taken in a KROS-1 camera, using the URS-60 type of apparatus with a BSV2-Cu tube, a 0.8 diaphragm, and RT-1 film. We studied bars 25 mm in diameter and up to 250 mm long prepared from an alloy of Zr–75%Nb.† For convenience of x-raying such long bars the standard sample holder of the KRSO-1 camera was supplemented with a

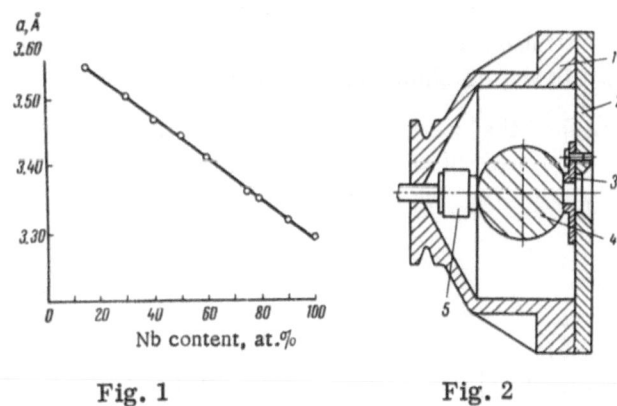

Fig. 1   Fig. 2

Fig. 1. Crystal lattice constant of the β phase of Zr–Nb alloys as a function of composition.

Fig. 2. Arrangement for fixing the sample in the holder of the KROS-1 camera. 1) Body of the holder; 2) top of the holder; 3) bushing; 4) sample bar (ingot); 5) spring device (clamp).

---

* Alloys containing over 20% Nb are of interest as superconductors.

† Here and subsequently the percentages are given in wt.% and the composition is based on the original mixture (charge).

special apertured disc (Fig. 2). A flat section etched in acid, 5 to 6 mm wide, was prepared along the whole length of the bar (see 4 in Fig. 2). The sample was pressed with the plane of the microsection against the plane of the disc; this enabled us to establish without ambiguity the distance from the sample surface irradiated by the primary beam to the film. This distance equalled approximately 40 mm. In order to determine the true distance we used an aluminum sample. The photography was carried out on a single film in a cassette with sector-shaped cuts. The exposure was 1.5 h. In order to ensure that the line on the x-ray diffraction picture should be continuous, the cassette carrying the film was rotated during the exposure, since the rotation of such a large sample in the KROS-1 camera was quite impossible.

The lattice constants of the β solid solution were determined by reference to the (330) line. The large Bragg angle ($\theta > 78°$ for the K_α radiation of copper) ensured an accuracy of ± 0.001 kxu in determining the parameter a, which was quite sufficient for establishing the composition of the alloy and revealing inhomogeneities to an accuracy of ± 0.35 Nb. The linear variation in the composition of the bars tested (Zr–75% Nb) is shown in Fig. 3. The maximum difference of composition in different parts of a bar produced by electron-beam melting was about 10%; almost always after the electron-beam melting of Zr–Nb alloys the zone with the greatest Nb content appeared in the head part of the ingot. The results obtained agree closely with chemical analysis, according to which the niobium content in the middle of the bar is 75% and at the end of the head part 80%.

The niobium distribution in a bar prepared by double electron-beam melting was also very nonuniform. There was a sharp fluctuation in niobium content in individual parts of the bar (from 71 to 79%). This was evidently one of the reasons for the serious degradation of current in wire obtained from such bars: 13 to 15 A in a solenoid as compared with 26 to 30 A in a short sample.

The use of arc melting for the primary production of the alloy improved the distribution of niobium in the bar; the deviation from the nominal content averaged about 1.5%.

The x-ray method may also be used for revealing inhomogeneities of composition in forged Zr–Nb bars. The variation in the composition of the β solid solution along a hot-forged Zr–75%Nb bar studied in this way is shown in Fig. 3d.

Thus we have established that x-ray analysis gives a more complete picture of the variation in composition along the length of a bar (and of the width of the zones in which the composition deviates seriously from its specified value) than chemical analysis.

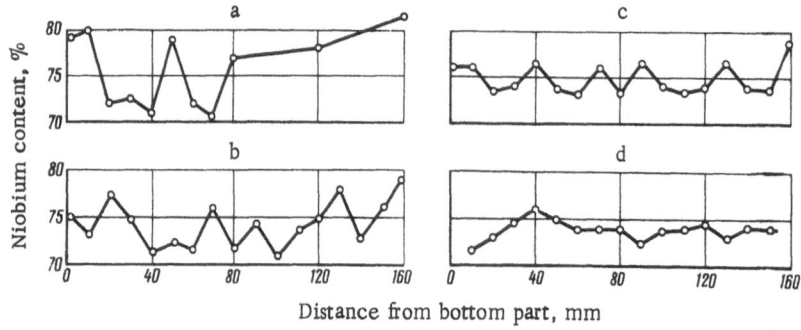

Fig. 3. Variation in niobium content along the bars. a) Electron-beam melting; b) double electron-beam melting; c) alloy melted in an arc furnace and remelted in an electron beam; d) hot-forged Zr–75%Nb bar.

By monitoring the quality of Zr–Nb bars in this way better statistics are secured and the analysis is faster and easier. In view of the fact that the local niobium content at a particular point of the bar or wire is determined by x-ray analysis, the x-ray method may be used to reveal very fine inhomogeneities in the alloy.

Checking the Homogeneity of Zr–Nb Wire

During the heat treatment of superconducting Zr–Nb wire, even that made from a high-quality bar of uniform composition, structural inhomogeneities may arise, for example, as a result of nonuniform thermal conditions. The revelation of inhomogeneities in the wire is extremely useful when developing the technology of heat treatment, selecting high-quality wire for the manufacture of solenoids, and so on.

The properties to be measured in order to monitor the homogeneity of the wire without damaging the latter must depend sharply on structure. Earlier studies of Zr–Nb alloys showed that these requirements were best satisfied by such physical properties as the thermo-emf (temf) and specific electrical resistance (ρ) at liquid-nitrogen temperature. For monitoring superconducting wire, the direct determination of the variation in superconducting characteristics along the length of the wire is particularly interesting; this may be done by the successive measurement of the critical current along the wire in a specified external magnetic field, as described by M. G. Kremlev et al. [1]. However, this method requires a supply of liquid helium and special equipment.

The possibility of using the temf method for checking the homogeneity of Zr–Nb wire is based on the fact that the thermo-electric properties depend sharply on the character and quality of new phases present; the separation of the α phase or an increase in its content reduce the temf (even to negative values), while the appearance of the ω phase has the opposite effect. If the wire has not been subjected to heat treatment, then a change in the Nb content from 25 to 75% has comparatively little effect on the value of the temf; hence the thermoelectric method is best used for heat-treated wire containing up to 50% Nb.

It should be noted that this kind of temf variation was observed when the material was paired with copper and the temperature gradient in the sample was created by having one end at -196°C (contact junction in liquid nitrogen) and the other at room temperature. The use of liquid nitrogen to create the temperature gradient was convenient, since in this way the sample suffered neither oxidation nor additional transformations.

The dependence of the electrical conductivity on the composition of the alloy and the change in this parameter on decomposition of the β solid solution also indicate the possibility of using this physical property for checking the homogeneity of the wire; measurements are best made at liquid-nitrogen temperature, since in this case the component due to thermal vibrations will be smaller.

The arrangement for measuring the temf along a Zr–Nb wire is shown in Fig. 4. The wire is wound on to a receiving coil 12, passes through liquid nitrogen 9, sliding along a copper contact 8, which at the same time constitutes the cold junction of the thermocouple formed between the sample wire and the copper standard. The signal from this thermocouple passes to the measuring circuit. The temf was determined relative to copper and all the leads and contacts were made of copper in order to prevent the development of parasitic thermo-emf.

The arrangement for measuring the electrical resistance along the length of the wire is shown in Fig. 5.

In studying wires the temf and electrical resistance were measured under semicontinuous conditions, either every one or two revolutions of the counter or else in places in which there

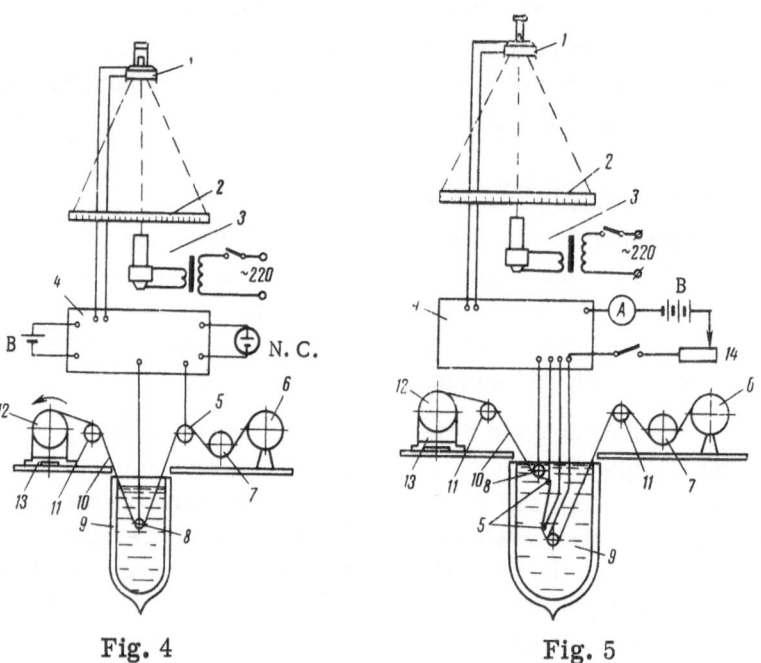

Fig. 4   Fig. 5

Fig. 4. Arrangement for measuring the temf along a wire. 1)
M17/2 mirror galvanometer; 2) scale; 3) illuminating system;
4) low-resistenace potentiometer of the PPTN-1 or R306 type;
5) hot contact; 6) reeling coil with wire; 7) mechanical rev.
counter; 8) cold contact; 9) liquid-nitrogen dewar; 10) wire
under test; 11) roller; 12) receiving coil; 13) electric motor
with reducer. B = dry battery; N. C. = normal cell.

Fig. 5. Arrangement for measuring the electrical resistance
along a wire. 4) Dc bridge of the R-329 type; 5) potential
contacts; 8) current contacts; 14) rheostat; A = ammeter. Remaining notation as in Fig. 4.

was a sharp deviation from the mean value. If the beam never passes off the scale or if auto-
matic-recording potentiometers (microvoltmeters) with corresponding scale limits are employed, the measurements may be made during the continuous motion of the wire. On using
devices with automatic recording the measuring is fed directly to the input of a single standard apparatus.

For the automatic recording of variations in R along the wire it is convenient to use a
differential measuring circuit.

Static measurements of temf and electrical resistance were also made on the wires under
consideration. In this case the wire was clamped in special contacts; after completing measurements on a particular part of the wire, the contacts were unscrewed and moved a few cen-
timeters along the wire. The results of the static measurements on long wires and short samples cut from these agreed closely with measurements based on a sliding contact.

The method of measuring the temf and electrical resistance was used in order to reveal
structural inhomogeneities along uncoppered wires (cleaned with aquadag) containing 25 and
30% Nb, the superconducting properties of which were determined earlier for the same samples.

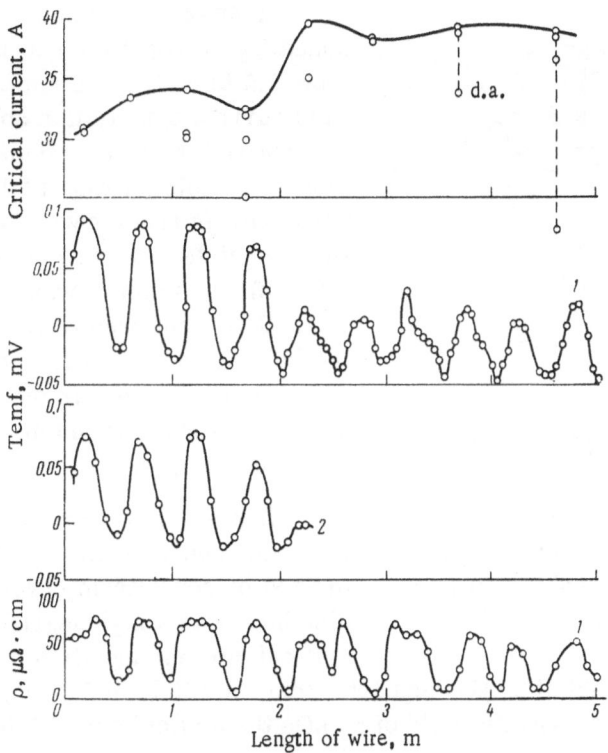

Fig. 6. Variation in critical current, temf, and electrical resistance along a Zr–25%Nb wire exhibiting a considerable conditioning (aging) effect (d.a. = results after aging). 1) Static measurements; 2) continuous measurements.

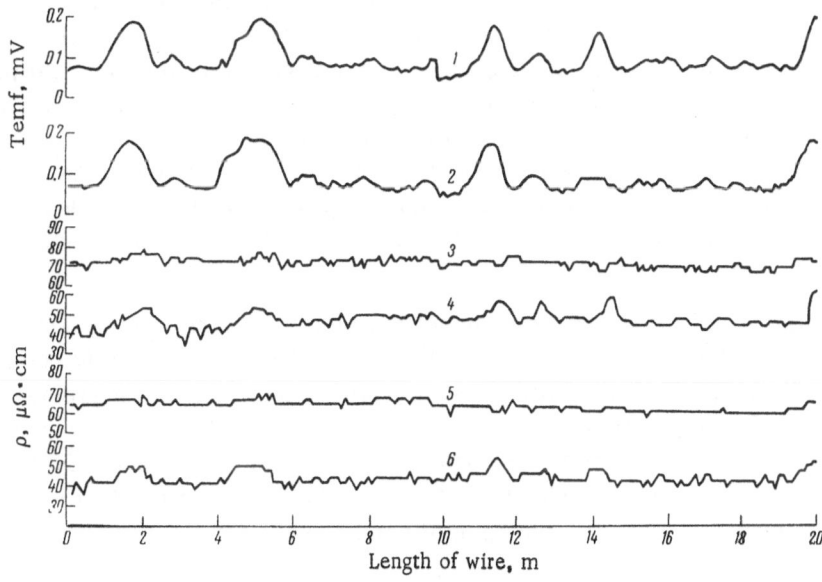

Fig. 7. Variation in temf and electrical resistance along a Zr–30%Nb wire with a high degradation factor. 1) Temf in static measurements; 2) the same in continuous measurements; 3) electrical resistance measured continuously at room temperature; 4) the same at liquid-nitrogen temperature; 5) electrical resistance, static measurements, room temperature; 6) the same, liquid-nitrogen temperature.

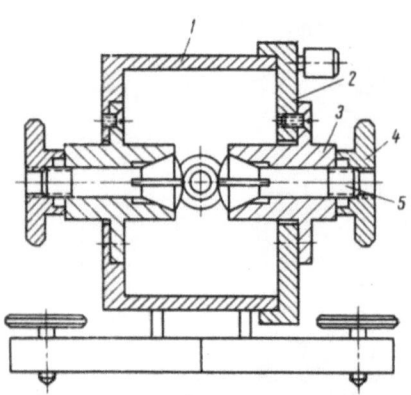

Fig. 8. Arrangement of the RKD camera with a special clamping device for the nondestructive study of wires. 1) Body of the camera; 2) top of the camera; 3) bushing; 4) nut; 5) clamp.

A Zr–25%Nb wire (subjected to intermediate annealing at 570°C for 1 h, and subsequent drawing from 0.5 to 0.25 mm diameter), having the same critical current density in a solenoid as in a short sample, showed the same values of temf and ρ (constant along its whole length) at 20°C and liquid nitrogen temperature. In another wire of the same composition (intermediate annealing at 500°C for 1 h, diameter 0.4 mm), for which it was known in advance that the critical current density varied along the wire and that the latter was subject to considerable conditioning* even for measurements on short samples, there was a periodic variation in temf and ρ of considerable magnitude (Fig. 6).

A Zr–30%Nb wire annealed at 500°C for 1 h and then reduced from 1 to 0.4 mm diameter, characterized by nonuniform physical properties (Fig. 7), also had quite a high degradation coefficient: The critical current density for short samples was over $2.5 \cdot 10^5$ A/cm^2 in a field of 30 kOe and 10^5 A/cm^2 in a field of 50 kOe, while in a solenoid with 930 turns situated in a magnetic field of 55 kOe the critical current density was $2 \cdot 10^4$ A/cm^2.

In order to discover the nature of the observed maxima and minima in the physical properties the phase and chemical compositions of the wire were determined by a nondestructive x-ray method. For this purpose a special attachment was made for the RKD camera; this had clamps designed to facilitate the fixing and centering of the wire under test (Fig. 8). During the exposure the wire was not rotated; however, since it had a fine-crystalline structure, the diffraction lines were quite continuous, although rather diffuse. The accuracy in determining the lattice constant was ± 0.01 kxu. X-ray analysis showed that the deviations in temf and ρ from the characteristic mean values were associated with specific variations in phase composition. In the case of the 30% Nb wire the maximum values of the physical properties (Fig. 7) appeared when only the β phase was present, thus explaining the high positive values of the temf; parts characterized by lower values (temf 0.07 to 0.08 mV) contained both α and β phases. A comparison of the line intensities showed that the amount of α phase was quite small; hence the temf remained positive. In the 25% Nb wire (Fig. 6) the α and β phases were found at both the maxima and minima of temf and ρ; however, in the latter case the amount of α phase was greater.

The nonuniformity of the phase composition along the wire is apparently associated with the fact that, during the aging process, the wire was not kept under isothermal conditions; in the parts heated to higher temperatures decomposition took place to a greater extent. The strict periodicity in the positions of the maxima of the properties being measured may be explained by the fact that the wires were wound on reels of diameter about 16 cm. It follows from this in particular that great care should be taken in the heat treatment of a superconducting wire to ensure the maintenance of isothermal conditions.

---

* The measurements were made by I. N. Goncharov in the United Institute of Nuclear Studies (Dubna).

Conclusion

The varying character of the decomposition of the solid solution along a wire made from original material of nonuniform composition and the existence of temperature gradients in the heat-treatment furnaces constitute two fundamental causes of variations in superconducting properties, even within the same batch of wire. In order to eliminate these variations, the technology of preparing and heat-treating the wires must be substantially improved. In order to monitor the quality of the product, the methods which have just been described for revealing inhomogeneities in the composition and structure of Zr—Nb wires and bars should be employed.

X-ray analysis may most conveniently be used for checking the homogeneity of bars rather than wires, since the number of analyses will be much smaller and less time will be wasted in preparing wire from inhomogeneous bars.

The homogeneity of uncoppered, heat-treated Zr—Nb wire may very conveniently be monitored by determining the temf and electrical resistance at liquid-nitrogen temperature. These properties may be measured rapidly, simply, and with a high accuracy, thus facilitating both selective and continuous monitoring of the wire.

The thermoelectric method of testing is more sensitive than that of measuring the resistance.

LITERATURE CITED

1. M. G. Kremlev, B. N. Samoilov, and S. S. Skulachenko, Cryogenics, 5(2):73 (1965).
2. T. G. Berlincourt and R. R. Hake, Phys. Rev. Lett., 9(7):293 (1962).
3. G. V. Zakharov, I. A. Popov, et al., Niobium and Its Alloys [in Russian], Metallurgizdat (1961).

APPARATUS FOR MEASURING THE CRITICAL CURRENT OF SUPERCONDUCTING METAL AND ALLOY WIRE SAMPLES IN AN EXTERNAL MAGNETIC FIELD

V. V. Baron and T. F. Demidenko

An apparatus for measuring I_K in a transverse magnetic field at 4.2°K is described. The measurements are carried out in a "frozen" field, which eliminates the necessity of continuous supplying the solenoid with current and reduces the amount of helium evaporating as a result of the evolution of Joule heat in nonsuperconducting current leads

In the manufacture of superconducting solenoids, in order to determine the prospects of using particular materials as windings, it is important to measure the parameters of the latter: the superconducting transformation temperature, the critical magnetic field, and the critical current.

The value of the critical current of a superconductor with fairly high T_K and H_K (for a given value of the magnetic field) determines the suitability of its use in solenoids.

It is already known that the critical current depends sharply on a number of metallurgical factors. However, at the present time it appears impossible to calculate the critical current of nonideal superconductors of the second kind for various values of magnetic field.

In order to determine the dependence of the critical current on the magnetic field strength in selecting superconducting materials, developing new compositions, and devising methods of mechanical and heat treatment, one tends to use special forms of apparatus consisting of a magnet or solenoid (usually superconducting, situated in a cryostat) and special devices for current supply and interlocking.

The critical current of the samples placed in the solenoid is determined by the four-contact resistance method. When studying the dependence of the critical current on the external magnetic field successive measurements of the critical current of the samples are made after reaching the required value of the field in the solenoid.

The efficiency of the apparatus and the consumption of liquid helium in making the measurements depend on the number of samples placed simultaneously in the apparatus and the number of measurements for various values of the field. Increasing the number of samples greatly increases the consumption of liquid helium owing to the greater number of current and voltage leads thus employed. Hence the number of samples measured in each filling of helium has to be limited (to between three and six).

In addition to this, for measuring the critical current in various magnetic fields one requires either a constant current supply to the solenoid during the experiments or the use of a superconducting "bridge" (cross-piece or connector) having a heater with which to feed the

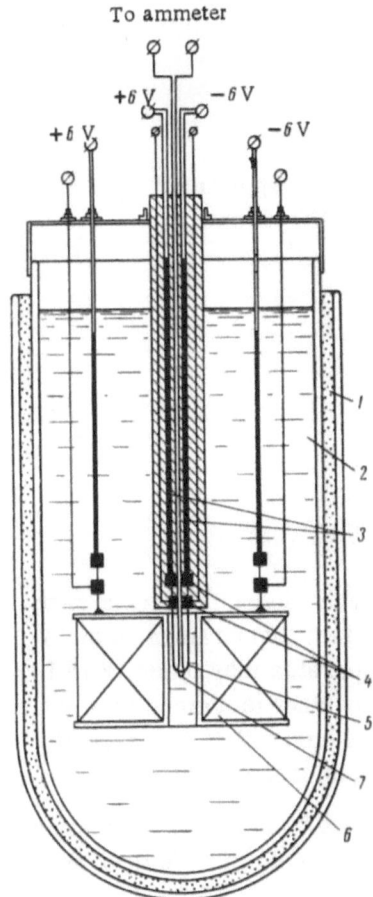

Fig. 1. Arrangement of apparatus with removable samples for measuring the critical current as a function of the magnetic field at 4.2°K. 1) Liquid nitrogen; 2) liquid helium; 3) lead conduit; 4) contacts; 5) sample; 6) superconducting solenoid; 7) bismuth magnetic-field meter.

solenoid and subsequently "freeze" the field. The number of these operations is determined by the required number of measurements at various field values. In both cases this also involves a considerable expenditure of helium.

In order to increase the number of samples measured for one filling of liquid helium and also eliminate the necessity of continuously supplying current to the solenoid creating the magnetic field (or incorporating a heater for the superconducting "bridge") we developed a method of measuring $I_K(H)$ differing from earlier methods in that, after the maximum magnetic field had been reached in the center of the superconducting solenoid, the field was "frozen" and calibrated with respect to height, so as to establish fixed values of field for various distances from the center to the edges.

The sample to be measured is placed in a thin-walled tube made of stainless steel, which is introduced into the working space of the solenoid after the "freezing" of the field. The samples are bent in the form of a hairpin so as to make the measured part perpendicular to the magnetic lines of force. This arrangement is necessary in order to reproduce the conditions which will be experienced when using the wire in solenoids in the course of the measurements. The current and potential leads connected with a relay of the RP-4 type (this operates when a drop of over 50 mV occurs in the samples) are also sited within the stainless steel tube. The arrangement is shown in Fig. 1a. A general view of the apparatus on filling with helium is presented in Fig. 2.

In view of the considerable change in magnetic field over the height of the solenoid for the ordinary form of winding and the comparatively small length-to-diameter ratio, the critical current/magnetic field relationship may be measured in the apparatus under consideration without having to "freeze" the field in order to secure various field values.

Thus, for example, on moving 50 cm up from the center of the solenoid (at which the field was a maximum and in the present case equalled 30 kOe) the field fell to 7 kOe. Lower values of field (from 7 kOe to zero) were obtained by moving 40 mm from the top of the solenoid.

By placing the sample to be measured successively in previously-determined fixed positions, the critical current/magnetic field characteristics could easily be secured.

The calibration and measurement of the field were carried out by the ballistic method using a bismuth measuring device. The great advantage of this apparatus is the possibility of easily interchanging the samples to be measured. The interchange is effected by removing the sample together with the central stainless steel tube and replacing the latter in the solenoid with a prepared tube containing the next sample.

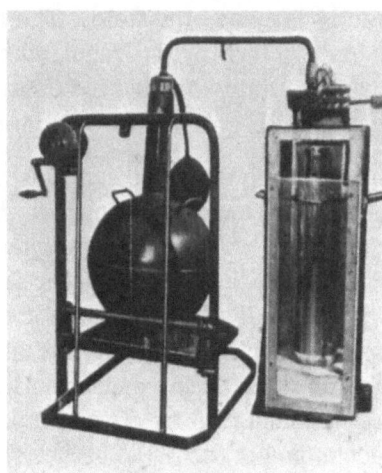

Fig. 2. General view of the apparatus
for measuring the critical current.

In case of need, an additional supply of liquid helium may be introduced during the measurements. The consumption of helium is not very great, since helium temperature is maintained in the cryostat. The number of samples measured in the apparatus may reach 12 to 15 with the consumption of about 10 liters of liquid helium. If a poor contact is noticed between the sample and the current leads (no field dependence of the critical current) it is replaced with another and a second measurement is made after the contact has been corrected.

If nondemountable samples are used, however, this requires the dismantling of the apparatus and the repetition of the experiment after the liquid nitrogen and helium have been renewed.